Cristina-Maria Pop

Non-equilibrium relaxation

Cristina-Maria Pop

Non-equilibrium relaxation

From language change to semiflexible polymer networks

Südwestdeutscher Verlag für Hochschulschriften

Impressum / Imprint

Bibliografische Information der Deutschen Nationalbibliothek: Die Deutsche Nationalbibliothek verzeichnet diese Publikation in der Deutschen Nationalbibliografie; detaillierte bibliografische Daten sind im Internet über http://dnb.d-nb.de abrufbar.

Alle in diesem Buch genannten Marken und Produktnamen unterliegen warenzeichen-, marken- oder patentrechtlichem Schutz bzw. sind Warenzeichen oder eingetragene Warenzeichen der jeweiligen Inhaber. Die Wiedergabe von Marken, Produktnamen, Gebrauchsnamen, Handelsnamen, Warenbezeichnungen u.s.w. in diesem Werk berechtigt auch ohne besondere Kennzeichnung nicht zu der Annahme, dass solche Namen im Sinne der Warenzeichen- und Markenschutzgesetzgebung als frei zu betrachten wären und daher von jedermann benutzt werden dürften.

Bibliographic information published by the Deutsche Nationalbibliothek: The Deutsche Nationalbibliothek lists this publication in the Deutsche Nationalbibliografie; detailed bibliographic data are available in the Internet at http://dnb.d-nb.de.

Any brand names and product names mentioned in this book are subject to trademark, brand or patent protection and are trademarks or registered trademarks of their respective holders. The use of brand names, product names, common names, trade names, product descriptions etc. even without a particular marking in this works is in no way to be construed to mean that such names may be regarded as unrestricted in respect of trademark and brand protection legislation and could thus be used by anyone.

Coverbild / Cover image: www.ingimage.com

Verlag / Publisher:
Südwestdeutscher Verlag für Hochschulschriften
ist ein Imprint der / is a trademark of
OmniScriptum GmbH & Co. KG
Heinrich-Böcking-Str. 6-8, 66121 Saarbrücken, Deutschland / Germany
Email: info@svh-verlag.de

Herstellung: siehe letzte Seite /
Printed at: see last page
ISBN: 978-3-8381-3203-7

Zugl. / Approved by: München, LMU, Diss., 2012

Copyright © 2013 OmniScriptum GmbH & Co. KG
Alle Rechte vorbehalten. / All rights reserved. Saarbrücken 2013

Contents

1 **Introduction**   1

2 **Language dynamics**   4
   2.1 Language evolution 5
       2.1.1 Sociobiological approach 6
       2.1.2 Sociocultural approach 10
   2.2 Language competition 17
   2.3 Concluding remarks 22

3 **Utterance Selection Model of language change**   25
   3.1 Single speaker dynamics 30
       3.1.1 Unbiased token production 32
       3.1.2 Biased token production 34
   3.2 Dynamics of N speakers 36
   3.3 Monte Carlo simulations 41
       3.3.1 Unbiased production 41
       3.3.2 The role of network topology 46
       3.3.3 Diffusion on a square lattice 47
       3.3.4 Biased production 48
   3.4 Utterance Selection Model on a mobile phone network 54
       3.4.1 Quasi-stationary frequency distribution 55
       3.4.2 Outlook 60

Contents

4 Language change in a multiple group society — 62
    4.1 Multiple Group Utterance Selection Model — 63
    4.2 Two groups — 64
        4.2.1 Weak coupling — 70
        4.2.2 Strong coupling — 73
    4.3 Many groups — 78
        4.3.1 Well-mixed system — 80
        4.3.2 Groups on a lattice — 87
    4.4 Conclusions — 89

5 Viscoelasticity of semiflexible polymer networks with transient cross-links — 93
    5.1 Single polymer dynamics — 95
        5.1.1 The worm-like chain model — 95
        5.1.2 Stochastic equations of motion — 97
        5.1.3 Mean square displacement and response function — 101
    5.2 Affine deformation in a polymer network — 105
        5.2.1 Mapping of N_0 filaments with constant force onto one polymer with varying force — 107
    5.3 Conclusions — 115

A Generating uncorrelated scale free random networks — 118

B Gillespie algorithm — 120

C Calculating the real part of the response function — 123

Bibliography — 127

1 Introduction

Language is one of the most prominent examples of complex systems. It is one of the basic tools of humanity, and yet many of its aspects still puzzle a broad range of scientists. The phenomena underlying the emergence and evolution of language, as well as cultural change, have been subject to increased interest from the physics community over the past two decades. We begin this work by discussing a number of mathematical and computational models of language dynamics.

Confining our attention to the ongoing processes acting on vocabulary, we only need to monitor our own use of language as it changes over time to see that words emerge, modify their shape and disappear, and sometimes even alter their meaning or functionality. The reasons for these changes are diverse. In order to transmit a message in as efficient and unambiguous a way as possible, a speaker will tend to use the conventions of her language. However, due to the fact that language not only communicates meaning, but does this in a way that reflects the speakers' cultural and social background, different linguistic variants ("different ways of saying the same thing" [1]) are used depending on the situation in which the speakers find themselves [2]. Also, because language "needs to keep pace with new realities, new technologies and new ideas, from ploughs to laser printers, and from political-correctness to sms-texting" [3], new means of expressing an idea can enter the language of a community via innovations from its members. Thus linguistic variants enter a competition for speakers. It is this aspect in the change of languages that we have a closer

1. Introduction

look at in the third chapter, by discussing various implications of the Utterance Selection Model of language change [1]. We also study this model on a network constructed from a mobile phone call data base as a surrogate of a real-life social network.

If speakers coming from distinct backgrounds find themselves united in a group, in time they will develop a common vocabulary in order to communicate successfully [4]. As our society consists of many groups, defined either by geographical location, age, profession or other criteria, we notice two antithetic tendencies that dominate the dynamics of the language: on the one hand, speakers affiliated with a social group will try to reach consensus on a variant in order to describe a particular situation. On the other hand, since this variant can differ from group to group, an element of rivalry between various forms stems from the interactions between speakers belonging to distinct social groups. In the fourth chapter of this work, our aim is to understand how the competition among word variants is resolved in such a society composed of several groups of speakers connected with each other, and how the strength of the connection between the groups influences the dynamics. To this end we want to find out how long it takes, on average, until the system relaxes to a state where only one variant is being used throughout the speech community, and which conditions have to be met for consensus to be a realistic outcome. We obtain scaling laws with nontrivial dependence on the parameters of the model and thereby demonstrate the importance of group structure.

Another ubiquitous example of complex systems are semiflexible polymer networks like the cytoskeleton, which is responsible for the structure and stability of living cells, as well as for their motility. In the final part of this work, we study the linear response to an applied macroscopic shear strain of a transiently cross-linked actin network, in which the filaments are modeled as *wormlike chains*. The binding kinetics of the transient cross-linkers provides a stress relaxation mechanism that has been shown experimentally to seriously alter the viscoelastic properties of the network. Our analytic approach serves the purpose of finding out how much of this effect is visible in the regime of low applied forces.

2 Language dynamics

Looking at the dynamics of language from the perspective of a natural scientist implies, beside making use of experimental data (which in this case means linguistic field studies, text corpora and speech records), the reformulation of linguistic theories in a mathematical framework, making use of known results from fields like evolutionary biology [5] and statistical physics [6]. Attempting to characterize a system as complex as language invariably means that one has to make simplifying assumptions, and finding the right balance between a realistic representation of a certain linguistic feature and a mathematically tractable form can prove to be a difficult task. Since we do not know how language emerged in the first place and how long this process took [7], models of language formation cannot be tested as such. However, one can try to establish some conditions under which a coherent communication system could emerge, since some of the forces involved are still acting on language as it is evolving today. Among these prerequisites is the ability of recognizing recurrent patterns in linguistic samples, as well as assigning them a meaning and finally using them in the appropriate context, this being the first step in the extraction of abstract rules. Whether faculties like processing linguistic structures with a complex hierarchy are directly inherited genetically or rather derived from a more general cognitive ability is still subject to controversy, reflected in the various modeling approaches.

From a distinct point of view, language is not only a tool used to communicate meaning, but can also be an asset of economical and even political importance, if we consider areas all around the world where several languages coexist, thus preserving in a certain sense the identity of the groups using them. An important question in the study of language dynamics is therefore the "competition" of languages for speakers. Due to various incentives or restrictions, speakers learn new languages and sometimes stop using the old ones, and in the long run there is even the possibility that the entire population will switch from its initial language to one better suited to its current needs. The perceived status of a language, depending on the social and economic benefits enjoyed by its speakers, as well as the number of speakers using it, are crucial factors in determining whether a language is endangered. However, it is not entirely understood how small groups can maintain their language for long periods of time, even though surrounded by a much more numerous community, as observed in various situations [8].

An account of the insight gained from the study of various mathematical and computational models can be found in [9, 10, 11, 12, 13, 14, 15]. In this chapter we will discuss a series of models of grammar formation, language change and competition between languages, from various points of view concerning the innate abilities of the individuals and the role of the interactions they are engaged in. The review of Castellano et al. [9] was used as a starting point for the account given in this chapter.

2.1 Language evolution

There are two schools of thought attempting to explain how language developed. The sociobiological approach rests on the hypothesis that successful communicators enjoy a reproductive advantage, and transmit their strategy genetically across generations. The sociocultural approach views language as a self-organizing system emerging from interactions between agents and the

2. Language dynamics

need of constructing a shared communication system. As Castellano et al. [9] remark, the dispute between these two communities goes along the lines of the debate "nature versus nurture", the innate qualities of individuals (nature) are weighed against personal experience (nurture) in determining differences in physical and behavioral traits. In the following we will have a closer look at these two directions and some of the mathematical models they gave rise to.

2.1.1 Sociobiological approach

The aim of this line of argumentation [16] is to explain the emergence of a coherent communication system by ascribing individuals endowed with high communication abilities a selective advantage over worse communicators. This is because successful communication is essential for cooperation, which increases the survival and reproduction probability of an individual. A main assumption, following the nativist approach of Chomsky (outlined in [17]), is that communication strategies are innate, and are thus transmitted genetically to the offspring. The *poverty of stimulus* argument states that children are exposed to a finite sample of language, but still end up speaking the same language. This is only possible if they choose from a finite set of candidate grammars, which are expressions of a *Universal Grammar* (UG). In order to achieve this, a *Language Acquisition Device* is needed, a set of internal rules on how knowledge is to be acquired systematically. Language change is in this context a result of the interactions between the UG and environmental conditions [18].

In [19, 20, 21], various strategies for learning the lexicon of a language are described. They comprise the development of a transmission and a reception mechanism (where the two are not necessarily correlated) from the communication samples that the individuals are exposed to. The formalism for these processes, based on evolutionary game theory, is presented by Nowak et al. [21, 22], relying on ideas formulated by Hurford [19]. Here, a population of individuals can produce a number of m signals to refer to n objects. Each

2.1 Language evolution

individual is characterized by two matrices, P and Q, called the transmission and reception matrices. These define the language L of the individual, in other words its behavior as speaker and listener, respectively. The entries p_{ij} of the $n \times m$ matrix P represent the probabilities that for describing object i, the individual will use the signal j. The entries of the $m \times n$ matrix Q, q_{ji}, denote the probabilities that a listener will associate signal j with object i. The rows of the two matrices are normalized:

$$\sum_{j=1}^{m} p_{ij} = \sum_{i=1}^{n} q_{ji} = 1. \tag{2.1}$$

In a typical communication scenario, an individual I_1 using language L_1 wants to transmit information about an object i to another individual I_2 who uses language L_2. For this, he will use signal j with probability $p_{ij}^{(1)}$. The listener will infer object i with probability $\sum_{j=1}^{m} p_{ij}^{(1)} q_{ji}^{(2)}$. One can obtain a measure of I_1's ability to convey information to I_2 by summing over all n objects: $\sum_{i=1}^{n} \sum_{j=1}^{m} p_{ij}^{(1)} q_{ji}^{(2)}$. When both I_1 and I_2 communicate information to each other, the success of the interaction can be expressed as a payoff function:

$$F(L_1, L_2) = \frac{1}{2} \sum_{i=1}^{n} \sum_{j=1}^{m} \left(p_{ij}^{(1)} q_{ji}^{(2)} + p_{ij}^{(2)} q_{ji}^{(1)} \right). \tag{2.2}$$

As can be derived from Eq. (2.2), the highest payoff is obtained when P is a binary matrix with at least one entry equal to 1 in every column (if $n \geq m$) or every row (if $n \leq m$), and Q a binary matrix with $q_{ij} = 1$ if p_{ij} is the largest entry in a column of P. For $n = m$, i.e. the number of objects equals the number of signals, the optimal solution is a matrix P with one 1 in every row and column, and the matrix Q the transposed matrix of P. The maximal payoff that can be obtained is $F_{\max} = \min\{m, n\}$.

In the transient period of language learning, when individuals are exposed to samples of language, they construct an association matrix A, an $n \times m$

2. Language dynamics

matrix whose entries keep count of how many times I_1 has heard object i being referred to with signal j by I_2, out of a total of k samples. The active and passive matrices P and Q are then derived from this matrix by normalizing the rows and columns:

$$p_{ij} = a_{ij} \bigg/ \sum_{l=1}^{m} a_{il}, \quad q_{ji} = a_{ji} \bigg/ \sum_{l=1}^{n} a_{lj}. \tag{2.3}$$

If the sample number k is very large ($k \to \infty$), A_1 will be the transmission matrix P_2. If $k = 1$, the association matrix will be binary, and therefore already normalized, so $P_2 = A_2$. Since the sampling is finite, language learning is a probabilistic process. If there are no errors in learning, eventually an absorbing state is reached. By definition (from the normalization of the transmission and reception probabilities), in this model one signal might describe several objects, thus homonymy is allowed, whereas several signals cannot always refer to the same object, synonymy therefore being excluded.

The population dynamics is as follows: having N individuals, one "generation" (meaning one step of the algorithm) ends after each of them interacts with every other individual and accumulates a total fitness

$$F_I = \sum_{J=1, J \neq I}^{N} (F_I, F_J). \tag{2.4}$$

This is equivalent to the reproductive fitness of the individual, the offspring being produced proportional to this payoff. Reproduction is asexual, meaning that each new individual has only one parent. The next generation learns the language using one of several strategies before replacing their parents' generation. These strategies define the peers for language sampling, which can be the parent, an individual with a high reputation or some randomly chosen individual. If the "teacher" belongs to one of the first two categories, the next

2.1 Language evolution

generation will perform better than they would if they learned from random individuals. Small errors in learning are favorable since they prevent the system from getting trapped in suboptimal situations where a signal is used to refer to several objects. Large errors however impair the optimization process and can lead to complex and unpredictable changes in language [23]. A way to reduce these learning errors is increasing the number of samples presented to the offspring. For the population to converge to a coherent language, there is a minimal number of samples that each individual has to be exposed to [24]. Another type of errors are the ones in perception [25]. Here a probability of misinterpreting a signal i for another signal j can be defined depending on the similarity between these sounds. In order to overcome the limitations that these errors impose on the number of objects that can be described with a given number of signals, sounds can be combined to form words, which can then be put together to produce sentences [26]. From this model it results that grammatical rules only evolve if the number of events to be described is larger than the number of words (see also [5, 27, 28, 29]). In addition to that, an upper limit for the number of words in an orally transmitted language can be found, if the distribution of words in this language follows Zipf's law[1] [30].

The model described above can also be formulated in terms of replicator-mutator equations [14, 31]. Again, we have an $n \times m$ matrix A, with entries a_{ij} which are nonzero if there is an association between object i and sound j. In the following, the matrix A is considered to be binary, therefore there are $M = 2^{nm}$ such matrices. Within a population of N individuals, a fraction x_k have association matrix A_k, with $\sum_{k=1}^{M} x_k = 1$. The evolution of the fraction x_k can then be described through the equation

---

[1]This law states that the frequency of a word is inversely proportional to the rank it occupies in a list of words ordered by decreasing frequency, i.e. the highest ranking word is twice as frequent as the second, three times as frequent as the third etc.

2. Language dynamics

$$\dot{x}_k = \sum_l f_l x_l Q_{lk} - \phi x_k, \quad l = 1, \ldots, M = 2^{nm}, \tag{2.5}$$

where $f_l = \sum_k F(A_l, A_k) x_k$ represents the fitness of individuals having association matrix A_l. $\phi = \sum_l f_l x_l$ denotes the average fitness of the population, and Q_{lk} is the probability that someone learning the language from an individual with A_l will end up using A_k. In the equation above, the second term on the right hand side ensures the constant size of the population. This equation is similar to the quasi-species equation [32, 33], but in this case the fitness values are frequency-dependent. In the limit of error-free learning, the replicator equations [34], known from evolutionary game theory, are obtained. In the case of imperfect learning, the error rate has to be below a certain threshold for linguistic coherence to be ensured, and the accuracy of language learning can be enhanced by long learning periods. However, learning involves certain costs which affect the fitness of the individual, so accuracy and learning speed have to be balanced against each other. It can be shown that there is an evolutionarily stable finite number of sampling events [35] that ensure a coherent language. In finite populations, accuracy of learning has to be higher than in infinite populations to maintain coherence [36]. A version of the model where individuals can change strategies is described in [37].

2.1.2 Sociocultural approach

In this class of models, a communication system that constantly evolves and self-organizes emerges due to simple interactions between agents. Good strategies do not ensure a higher biological fitness, but only better communication abilities. New rules crystallize due to collective action rather than inherent mechanisms. This framework is more flexible regarding innovations, agents being endowed with inventing abilities.

A paradigmatic model for this category, the Naming Game formulated by

2.1 Language evolution

Steels [38], shows how complex self-organization can emerge from simple rules. Agents start developing their own vocabularies in a random fashion. In the course of interactions with other agents they exert a number of tasks like perceiving a particular object, drawing the interlocutor's attention to an object, ascribing meaning to an utterance. Afterwards they adjust their vocabularies according to the response of the other agent. The more successful communication is, the less the language changes, and thus a shared vocabulary evolves. A similar model is the one of Ke et al. [39], which describes the formation of a common vocabulary through simple local interactions, where the agents imitate either a random agent they interact with, or the behavior of the majority. Their interaction model, like the one presented by Lenaerts et al. [40], uses a probabilistic description of vocabulary, similar to the evolutionary game theory approach. However, in this case the matrices are changing after each interaction: if communication is successful, i.e. if the listener ascribes to the received signal the meaning that the speaker intended to convey, the corresponding entry in the speaker's production and the listener's reception matrix is increased by an amount Δ. Due to normalization, the other entries of the two matrices are decreased. This means that in further interactions, the speaker will rather use a signal that she has produced before when referring to a certain meaning, whereas a listener will be more likely to relate a heard signal to a meaning that has been expressed in the same way before.

A minimal version of the Naming Game introduced by Baronchelli et al. [41] comprises N agents on a fully connected network, trying to establish a common name for a given object. Every agent has an inventory of words for referring to the object, which is empty at the beginning. For each interaction, two players are chosen at random, one of them acting as speaker and the other one as listener. The speaker chooses a word from her[2] inventory, or, if this is empty,

---

[2] Throughout this work we will use the convention that when referring to the speaker, the female pronoun is used, whereas for the hearer we employ the masculine.

2. Language dynamics

she invents a word. If the listener knows the word, communication is successful, and both erase from their inventories all words except for the one just used. If the listener does not know the word, he adds it to his inventory, and the interaction is classified as a failure (Figure 2.1[3]).

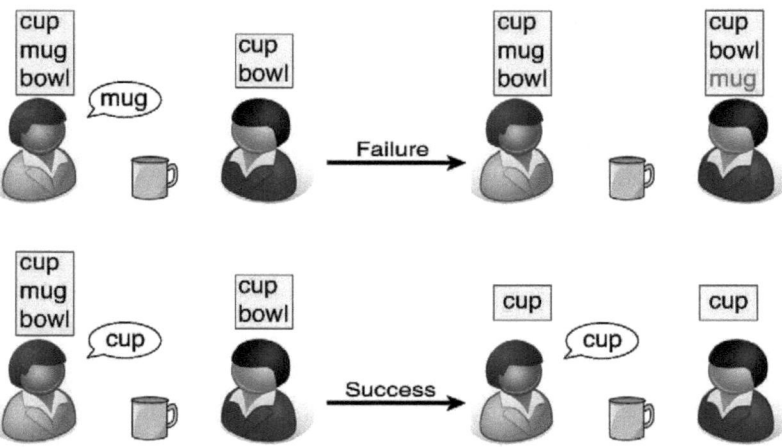

Figure 2.1: An interaction step in a minimal version of the Naming Game. If the listener does not have the uttered word in his inventory, he adds it, and the interaction is considered to be a failure (top). If he knows the word, both agents erase every other word from their inventories, keeping only the one that has been uttered by the speaker and recognized by the listener. The conversation is then considered to be successful (bottom). In the beginning, the agents' inventories are empty, so when they act as speakers for the first time, they invent a word.

The parameters of interest are the total number N_w of words known by the population (the sum of all words known by each speaker), the number of different words N_d, and the success rate $S(t)$. In the initial phase, the number of words increases, up to a certain point when order starts emerging spontaneously. The system undergoes a disorder-order transition to an asymptotic

---
[3]This figure, as well as some of the figures in Chapters 3 and 4, was created using templates provided by Office.com

2.1 Language evolution

state of consensus, where the same word is used by all agents. Power laws as a function of the system size are found for the convergence time and the maximal number of words. The influence of the network topology on the outcome of this minimal Naming Game is discussed by Dall'Asta et al. [42].

Considering an even simpler model with only two possible words, one can formulate a deterministic description for the fraction of agents knowing one of the two words, or both of them [9]. Depending on the initial conditions, one or the other word can take root in the system. The state where the initial fractions are equal is unstable, a perturbation would make the system reach one of the two states where one word dominates. An extension of this idea was treated by Blythe [43], a hybrid model unifying the Naming Game and the Voter Model, one of the most well-known models of opinion dynamics [9, 44]. Here also, the phase space displays two phases, depending on the parameters of the model: one in which consensus on one word is reached, and the other where, only in the deterministic limit, the two words coexist with equal frequency. Similar ideas are found in [45]. These models resemble language competition models, like the one of Abrams and Strogratz [46], which we will have a closer look at in the next section.

Starting from models of cultural transmission such as the one of Cavalli-Sforza et al. [47], Nettle [48] uses social impact theory to describe language change. In his model, individuals have a choice between two words, p and q. The agents are characterized by their age, which can take values from 1 to 5, hereby defining children, teenagers, young adults etc. They are placed on a two-dimensional lattice, each agent being influenced by the ones around him, this influence decreasing with distance. In each life stage, agents sample the language of the agents surrounding them, provided these are not aged 1, and the end of the stage, the agents grow one unit older. The individuals that were previously aged 5 die and are replaced by children. At the beginning, everyone uses word p. In the first two life stages, errors in learning can occur, opening

2. Language dynamics

a path for innovations. In this context, it means that the agent learns word q instead of p with a small probability. If the error happens in stage 1, there is a good chance that it will be corrected in the next life stage. Errors acquired in stage 2 however are preserved until death. From stage 3 on the learning process stops, and the only way that agents contribute to language evolution is by influencing younger speakers. One of the findings is that tuning the mutation rate within reasonable limits does not produce linguistic change, so word q will not impose itself through this mechanism only. To simulate situations in which the consensus variant changes, differences in status and functional bias are introduced. By drawing the status of the agents from a Poisson distribution, a society with a few very influential individuals is created, who can spread the rare variant in the community. Functional bias means there is some inherent bias against some variant, for instance because it is hard to learn. It appears that a small functional bias without social differences does not lead to linguistic change.

In a second paper [49], Nettle uses the same model to show that the rate of language change decreases as the speech community increases, eventually reaching a stable asymptote. Linguistic borrowing is simulated by placing in the community a few young adults (aged 3) with high status, who speak the less preferred variant. One can observe that small communities are more likely to borrow the variant of the immigrants than larger ones, if the number of foreigners is fixed. This does not hold however if the proportion of foreigners with high status is fixed. Finally, smaller communities seem to be a more fertile ground for structures against which there is a bias, here they can evolve and survive longer than in large communities.

Ke et al. [50] explore the model proposed by Nettle on various types of networks. They find that, when functional bias is sufficiently high, innovations spread linearly in regular and small-world networks, and very quickly, following an S-shaped curve, in random and scale-free networks. However, the success

2.1 Language evolution

rate of change is lower in the latter category. By introducing statistical learners who use both variants according to their frequency and impact in the sample they collect, a variant with a small functional advantage can diffuse easily.

Another approach coming from computational linguistics is the Iterated Learning Model (ILM) developed by Kirby et al. [51, 52]. Here, individuals infer rules from samples offered by a teacher, incorporate new rules into the grammar and then generalize these rules. Through this self-organizing mechanism, the *poverty of stimulus* issue can be overcome. Irregularities in languages can be accounted for by considering pressures on language, like choosing the shortest signal out of a set of equivalent alternatives, or failing to pronounce all sounds of a word (erosion).

Niyogi and Berwick [53] compare the ILM with Social Learning (SL), where individuals sample utterances of several teachers, for example two parents. They show that whereas the ILM only allows for linear dynamics, in the case of social learning the dynamics is nonlinear and thus displays a wider set of possible outcomes. Applying the SL model to a situation of two languages in contact, they find that it correctly predicts an observed phase transition that results in the loss of some types of syntactic constructions. With the linear ILM, this could not have been achieved.

Griffiths et al. [54, 55] incorporate Bayesian learners into the Iterated Learning paradigm of Kirby et al. These have some a priori expectations about the structure of language, which are expressed in a prior probability distribution. By sampling utterances of other users, they use Bayes's rule to infer the structure of language used in the community. The first agent to provide samples is assumed to have already learned the language. By making particular choices regarding the prior distribution, they find that the speakers' word usage frequencies are governed by neutral evolutionary dynamics. In [55] this theory is used to explain S-shaped curves in language change, power law distributions of word frequencies and the effect of frequency in word change rates.

2. Language dynamics

Croft [2] suggested an evolutionary language change approach, which was cast into a mathematical framework by Baxter et al. [1]. This model, which we will discuss in detail in Chapter 3, is based on utterance selection. This means that word variants replicate through speakers' utterances, and genetic drift is due to the stochasticity of utterance production.

A somewhat more "experimental" course in explaining changes in language was taken by Lieberman et al. [56] and Pagel et al. [57]. By studying corpora of texts spanning very long time periods, they show that the rate of word evolution is strongly related to the frequency of word use. Words used rarely are more susceptible to change, and Lieberman et al. motivate in this way the regularization of verbs in the English language. They find the half-life of irregular verbs to scale as the square root of the frequency with which they are used.

On shorter time scales, Altmann et al. [58] show that the *word niche* (quantifying the dissemination of words across speakers) is more important for the success of words than usage frequency when considering the dynamics of the whole vocabulary. By analyzing data from Internet discussion groups, they conclude that poor dissemination of words leads to a reduction of their usage frequency.

Ferrer and Solé [59, 60, 61, 62] investigate the validity of Zipf's law with the aid of large text corpora. They find two different scaling regimes [59] for a kernel lexicon consisting of common words and a large lexicon used for specific communications, respectively. Their conclusion is that the standard Zipf's law describes very accurately words that are known to a very large number of speakers. In [61], a mathematical framework based on a simple optimization process applied to a binary association matrix (as the ones discussed in the previous section) is employed to model effort minimization of both speaker and hearer. The findings show that the optimal solution for maximizing the transmission of information under constraints of speaker effort is precisely Zipf's law.

2.2 Language competition

There are roughly 6000 languages spoken today, and they are disappearing at a rate of two a month [63, 8]. This means that by the end of the 21th century, between half and three-quarters of today's languages will go extinct, most of them being the languages of very small, often isolated societies. There are many reasons which cause languages to disappear, among them colonization and economic globalization, but generally speaking one can say that languages are threatened by extinction in the same way as biological species [64, 65, 66]. In the attempt of understanding how this process takes course and which would be the crucial points where one could intervene, an avalanche of models has been initiated by the work of Abrams and Strogatz [46].

In models of language competition, language is mostly simplified to the extent that it has no intrinsic structure. There are two categories of models where various languages compete for speakers, one of them studying the evolution of macroscopic parameters such as fractions of the population, and the other the fate of individual agents.

We begin with discussing the model of Abrams and Strogatz [46], where speakers are monolingual in one of two possible languages and highly connected. A language becomes increasingly attractive with the number of speakers and its perceived status. The latter is a parameter describing social or economical opportunities that its speakers enjoy, as perceived by society. The probability per unit time of a speaker to change from language X to language Y is $P_{yx}(x, s)$, where x is the fraction of speakers using language X, and the relative status of language X, denoted by s, takes values in the interval $[0, 1]$. Then the speaker dynamics is given by the following equation:

$$\frac{dx}{dt} = yP_{yx}(x, s) - xP_{xy}(x, s), \qquad (2.6)$$

where $y = 1 - x$ is the fraction of speakers of language Y. If a language has status

2. Language dynamics

$s = 0$ or no speakers, it will never be used again. The transition probabilities are assumed to be of the form $P_{yx}(x,s) = cx^a s$ and $P_{xy}(x,s) = c(1-x)^a(1-s)$ respectively, with the exponent $a = 1.31 \pm 0.25$ obtained by fitting various data sets. From the analysis of the model it results that the coexistence of the two languages is not a stable solution, and thus one language will eventually drive the other one to extinction.

In [67], Patriarca and Leppänen enrich the model of Abrams and Strogatz by a geographical component. Generalizing Eq. (2.6) to a reaction-diffusion equation and dividing the total area in two "influence zones" where the status of one or the other language is higher, they find that the languages can coexist, each language being spoken mainly in its own influence area, and the two languages interacting at the interface. In [68], the population is allowed to grow. If there is no geographical barrier between the two languages, one language survives. Which of the two it is depends on the initial distribution of the speakers and the boundary conditions. In the presence of geographical inhomogeneities, the two languages can coexist in different areas, as in [67].

The approach of Pinasco and Romanelli [69] explains coexistence of languages in the setting of a Lotka-Volterra type [34] extension of the Abrams-Strogatz model. The conversion rates are proportional to xy rather that yx^a, which however does not result in qualitative changes of the dynamics. The carrying capacities in the absence of competition ($c = 0$), S_x and S_y, are determined by the environmental conditions. The equations for language competition are then

$$\frac{dx}{dt} = cxy + \alpha_x x \left(1 - \frac{x}{S_x}\right) \quad \text{and} \tag{2.7}$$

$$\frac{dy}{dt} = -cxy + \alpha_y y \left(1 - \frac{y}{S_y}\right), \tag{2.8}$$

where the coefficient c is the rate of conversion from y to x (i.e. the status of language x), and α_x and α_y are positive parameters including natality and

2.2 Language competition

mortality rates of each population. Eqs. (2.8) describe a combination of a predator-prey model, where the prey mutates into a predator after being captured (a "vampire-type" predator, as the authors call it), and an epidemics model, where infected individuals do not recover, but also do not die immediately. This model is asymmetric, in that one language is more prestigious and attracts speakers of the other language, but there is no incentive for its speakers to change to the other language. The non-attractive language can survive in the same area as the other if it fulfills certain conditions, like having a high growth rate and a low switching rate to the attractive language, and if the carrying capacity of the latter is small (and thus the speakers of the attractive language have to fight for resources).

Another extension of the Abrams-Strogatz model, allowing for speakers to become bilingual, was studied by Mira and Paredes [70]. They modify Eq. (2.6) to

$$\frac{dx}{dt} = yP_{yx} + bP_{bx} - x(P_{xy} + P_{xb}), \qquad (2.9)$$

where b denotes the fraction of bilingual speakers. They further include a parameter quantifying similarity between languages, k, which takes values between 0 and 1, with $k = 0$ meaning that monolingual speakers of a language cannot communicate with monolingual speakers of the other. In this context, the transition probabilities change to $P_{xb} = ck(1-s)(1-x)^a$ for monolingual speakers of X becoming bilingual, and $P_{xy} = c(1-k)s(1-x)^a$ for speakers completely changing from language X to Y. The probabilities for monolingual speakers of Y to become bilingual or switch to language X are $P_{yb} = cks(1-y)^a$ and $P_{yx} = c(1-k)s(1-y)^a$ respectively. The transition from bilingual to monolingual speakers is given by $P_{bx} = P_{yx}$ and $P_{by} = P_{yx}$. From studying these equations, one learns that if languages are similar enough, i.e. the parameter k is higher than a particular value (which depends on the relative status s), the two languages can coexist.

2. Language dynamics

Minett and Wang introduced bilingual speakers in a slightly different way in [13, 71]. Here, the two languages do not have to be similar to each other to survive, but the transmission rules are different. On the one hand, there is vertical transmission from parents to children, where children of monolingual parents inherit the only language of their parents, whereas children of bilingual parents can learn one or both languages. With rate μ, parents are replaced in this way by their children. With rate $1 - \mu$, the language transmission is horizontal, implying that monolingual adults can also learn the other language and thus become bilingual, or they can remain monolingual in their language, but they will never completely switch from one language to the other. The transition rates of the Abrams-Strogatz model are modified according to these rules. Their findings are that a state where both languages survive, be it through bilingual speakers or monolingual speakers of each language, is unstable, and thus one of the languages is prone to go extinct. Formulating their model in an agent-based setting, they find that in strongly connected networks, the extinction of a language can be prevented more easily than if connections are sparse, and the way of doing this is by increasing the status of the language when the fraction of speakers falls beyond a certain threshold. A further generalization of these ideas is presented in [72], where the status of the endangered language can be tuned to take values in a continuous interval, in order to preserve the coexistence of the two languages.

The Abrams-Strogatz model was also the starting point for numerous agent-based models. In [73], the differences between a mean-field and a stochastic version of the Abrams-Strogatz model are discussed, namely the role of fluctuations in the fixation of a language. In the microscopic model, coexistence of the two languages is never achieved, since fluctuations will always drive the system into one of the two absorbing states where only one language is spoken.

Castelló et al. [74] map the Abrams-Strogatz model on the Voter Model [44] with an additional state representing bilingual speakers. This bilingual

2.2 Language competition

state changes the coarsening dynamics of the Voter Model and speeds up the extinction of one of the languages. Both the Abrams-Strogatz model and the one including bilingual speakers, presented in [74] are studied in [75] and [76] on scale-free random networks, the results show that language coexistence is lost more easily in the model with bilingual individuals. Also, coexistence is more stable in strongly connected networks, as connectivity is decreased, one language dominates. In networks with community structure, the presence of bilingual individuals can help minority languages to survive for longer time periods.

The Schulze model [12] describes a society in which the language of each speaker has F independent features, each of them taking one out of Q values. In the case of $Q = 2$, language is a bit string, as in the quasi-species model [33]. Language can change through three processes: on the one hand, there can be spontaneous mutations in the value of a feature with probability p. There can be word transfer from one language to the other, this happening when a feature takes over the value of the same feature of a neighbor's language with probability q or a value at random otherwise. Finally, the agent can learn a new language: with probability $(1-x)^2 r$, the speaker switches to another language spoken by a fraction x of the population. Placing the agents on a network, only the immediate neighbors have an influence on a speaker's language. In a variant of this model [77], communities are allowed to grow, individuals can reproduce and die. For low spontaneous mutations (small p), the majority will speak one language, whereas for high values of p, many languages coexist and show a log-normal distribution. In [78], the authors find better agreement with real data by sampling the distribution over a long time interval including transient times.

Zanette [79] presents an analytical formulation of the Schulze model, which resembles the quasi-species models discussed in Section 2.1.1. Here, however, there is no intrinsic fitness of language, but the fitness grows with the spreading

2. Language dynamics

of the language. In [80], Zanette argues that the log-normal distribution of language sizes can be understood as a consequence of demographic growth.

In another work, Schulze and Stauffer analyze language switch in a region that has been conquered by speakers of a different language. On a Barabási-Albert network, the language of the conquerors always imposes itself, whereas on a two-dimensional lattice the time for this language to take over diverges.

Another type of language competition model is introduced by de Oliveira et al. [81]. Here, starting with an ancestral language, the population spreads from an initial site across a geographically inhomogeneous territory, modeled by a two-dimensional lattice with different fitness values per lattice site. The fitness of a language is given by the "friendliness" of the landscape, i.e. the amount of resources available on the occupied territory. The more unfriendly the environment is, the faster the language mutates, hereby giving birth to new languages. The number of languages grows as a power law with respect to the occupied area. As soon as the whole territory is occupied, the language creation process comes to a halt. The various different populations engage in a competition for territory, with the languages existing in friendlier environments having a higher fitness. In time, languages occupying a larger territory will gain more and more fitness, and in the end only very few languages survive.

2.3 Concluding remarks

When confronted with the complexity of language as we know it, we realize that in order for it to have evolved this far, the individuals using it need to be endowed with mental abilities like the understanding of temporal order, causality, locality or abstraction, apart from certain physical requirements like the lowering of the larynx [82, 83] that made it possible to produce sounds in the first place. Therefore, when considering the arguments of the sociobiological approach to language evolution, a certain amount of innate qualities can be

2.3 Concluding remarks

accounted for. In this sense, a language that can express statements with a higher degree of complexity has a higher fitness than a language that can only express one idea at the time by being able to convey more information in an efficient way, and therefore speakers will continue to refine the language as long as it reduces ambiguity. This way, one can imagine the emergence of syntax. However, the implication that the speakers of a "fit' language will all be successful reproducers is not entirely justified. That speakers should inherit the successful strategies from their ancestors is a very strong assumption, this approach leaving too little room for the features of the individual, and maybe the aspect of learning the language is not ascribed enough importance.

In this respect, the sociocultural approach leaves the most room for the creativity of agents, since structured communication emerges from interactions on the microscopic scale, without speakers or a language being endowed with a priori fitness. Language change can be studied on various levels and therefore on different timescales, from alterations in vocabulary, which happen in the course of months or few years, to the regularization of verbs, which has been going on for centuries.

Most models of language competition simplify language up to the point where it becomes just another feature of the speakers, like eye color or political preference, hereby losing all attributes that are specific to it. Estimating the parameters that are introduced, like the perceived status, or in some cases the carrying capacity of a language, is nontrivial in a real-life situation, so the insight gained from the analysis of various models is not necessary applicable to other situations. As Krauss [63] argues, it is very difficult to state when a language is dying, but even if one could predict from available data when a language will go extinct, there are many other external factors like politics, environment, etc., which can render the implementation of suggestions on how to prevent language death very complicated. Also, in most models, new languages are either not allowed for, or they can only emerge in an initial transient phase. There is a

2. Language dynamics

discrepancy between this constraint and the observation that new languages appear, for example, as a result of a difficult sociopolitical situation, as in the case of Serbo-Croatian, which gave birth to three distinct languages: Serbian, Croatian and Bosnian [8].

Since nowadays linguistic data is readily accessible even for non-linguists in the form of collections of scanned books spanning enormous periods of time, archives of publications from the last decades, as well as data from various social networks, it would be desirable to construct more "experiments" in order to enhance models until they withstand the confrontation with real data.

In the following chapters we will pursue the aim of capturing the effects of forces acting on short timescales that can be observed in the course of a human lifetime. We will focus on the changes experienced by the vocabulary of a language.

3 Utterance Selection Model of language change

If we were to study someone's vocabulary along the years, we would notice a lot of changes. Not only are words continually added but some of them fall into disuse for long periods of time, and many of them eventually disappear. The language use pattern of an individual varies due to many reasons like aging, a change of residence, media influence. The particular situation in which the speaker finds herself can have a strong influence on the choice of words [84]. The frequencies with which words are used are a way of quantifying language change on the word level. To get a general idea about how these change in time, one can explore text corpora available online with interfaces like Google Ngram Viewer [85]. In Figure 3.1 we show the change in frequency of the words *computer*, *PC* and *calculator* in a time span of 72 years. The plot only reflects the usage of these words in written language, since the text corpus behind this application consists of millions of books. We see that the curves experience an increase around the time when objects known under that name (or acronym) became popular in wide parts of the society.

From now on, we will use the term "language" to describe spoken language. This differs from written language insofar as it changes on a faster time scale, observable in the course of a human lifetime. The records of written language often do not capture the whole spectrum of changes in a language over shorter periods, the reason being that some of them are alterations in pronunciation,

3. Utterance Selection Model of language change

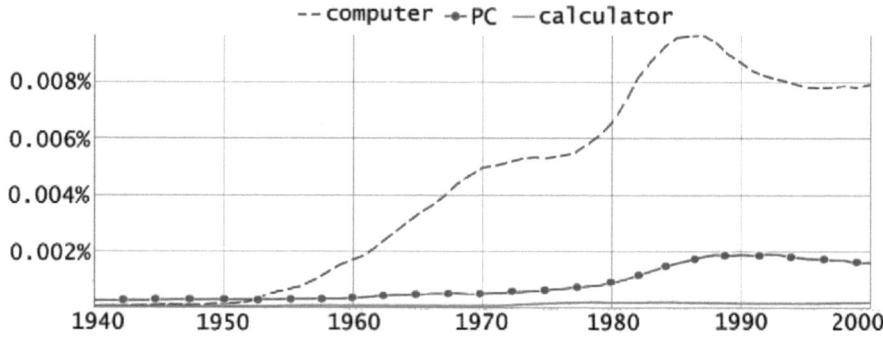

Figure 3.1: Google Ngram Viewer [86]: the evolution of the frequencies of the words *computer*, *PC* and *calculator* from 1940 to now, obtained from sampling books appeared in this period.

which take a long time until they become reflected in the way the words are written, and others are short-lived vogue words.

Baxter et al. [1] formulated an evolutionary model of language change based on the ideas presented by Croft [2]. It describes a very simple language, consisting of a single replicable structure called lingueme (the linguistic analogue of a gene, in the example presented in Figure 3.1 it is the *computer* concept), with V variant forms (the equivalent of alleles, in Figure 3.1 the words *computer*, *PC* and *calculator*). Selection steps in when speakers interact and utter these variants, hereby leading to the replication of the lingueme.

The underlying social network consists of N speakers who use these variants according to their knowledge of language, with certain frequencies, which are stored in the vectors $\mathbf{x}_i = (x_{i1}, x_{i2}, ..., x_{iV})$, with $i = 1, ..., N$. These frequencies are normalized to 1 for each speaker:

$$\sum_{v=1}^{V} x_{iv}(t) = 1, \quad \forall i, t. \tag{3.1}$$

The state of the system can be characterized at any time t through the entirety of agents' vocabularies $X(t) = (\mathbf{x}_1(t), ..., \mathbf{x}_N(t))$. The initial conditions can be chosen arbitrarily, giving the possibility to start with each speaker knowing only one variant or all speakers using several variants with different frequencies.

Figure 3.2: A social network for the Utterance Selection Model can have weighted links, if one wants to model heterogeneous communication patterns of speakers, or all links can have the same weight, which would give the speakers equal changes to interact.

The interaction algorithm between individuals relies on the following steps: first, two speakers are chosen at random, with the condition that they are connected by a link. The probability of choosing a particular pair corresponds to the weight of the link, G_{ij} (if the links have equal weights, then also the probabilities of choosing each of them for an interaction are equal). In the next step, each of the speakers produces a string of tokens of length T. The tokens are instances of the V variants, uttered according to frequencies x'_{iv}. The numbers $n_{i1}(t), ..., n_{iV}(t)$ of tokens of each variant produced by speaker i are drawn from the multinomial distribution

$$P(\mathbf{n}_i, \mathbf{x}_i) = \binom{T}{n_{i1} \cdots n_{iV}} (x'_{i1})^{n_{i1}} \cdots (x'_{iV})^{n_{iV}}, \quad (3.2)$$

where $\mathbf{x}_i$ is the frequency vector of speaker i and $\sum_{v=1}^{V} n_{iv} = T$. If the token production is unbiased, these frequencies are equal to the entries in the speaker's vocabulary vector, $x'_{iv} = x_{iv}$. If there is a bias in production, the probabilities

3. Utterance Selection Model of language change

to utter tokens of a particular variant are a linear transformation of the stored frequencies: $x'_{iv} = \sum_w M_{wv} x_{iw}(t)$, where the matrix M, same for all speakers, can be seen as the effect of universal forces such as articulatory constraints, according to [1], and its columns sum up to 1 so that the production frequencies are properly normalized. One of the effects of bias is that the speaker can also produce tokens of a variant which in her vocabulary has frequency zero. The case where $M_{wv} = 0 \; \forall v \neq w$ corresponds to no bias. After both speakers have uttered the tokens, their vocabularies are updated, taking into consideration the old entries of the frequency vectors, the utterances of the speaker herself, as well as those of her interlocutor:

$$\mathbf{x}_i(t + \delta t) = \frac{\mathbf{x}_i(t) + (\lambda/T)[\mathbf{n}_i(t) + H_{ij}\mathbf{n}_j(t)]}{1 + \lambda(1 + H_{ij})}. \qquad (3.3)$$

Figure 3.3: Interaction between two speakers: Here the number of variants $V = 2$ (cup and mug), and each speaker produces $T = 5$ tokens according to the frequencies she has stored (cartoon inspired by Figures 2-4 in [1]).

The update rule Eq. (3.3) ensures the correctness of normalization. The parameter λ describes the magnitude of the change occurring in the vocabulary of a speaker due to an interaction, whereas H_{ij} is the weight given by a speaker to her interlocutor's utterances relative to her own. The steps of the algorithm are repeated until either there is only one variant spoken in the community (for unbiased token production) or another stationary distribution is reached (for biased production).

In the limit $H_{ij} \to \infty$, one can consider an asymmetric version of the two-variant Utterance Selection Model where in an interaction each agent behaves either as a speaker or as a listener. If the speaker produces only one token, the dynamics corresponds to the Voter Model [43], one of the simplest models of opinion formation. In contrast, the symmetric Utterance Selection Model, which we are considering, exhibits a much richer dynamics stemming from each agent being both speaker and listener at the same time.

In order to understand the frequency distribution of the linguistic variants, in [1] the authors analyze the Utterance Selection Model in the continuous-time limit with the aid of a Fokker-Planck equation obtained via a Kramers-Moyal expansion [87]:

$$\frac{\partial P(\mathbf{x},t)}{\partial t} = -\sum_{v=1}^{V-1} \frac{\partial}{\partial x_v}\{\alpha_v(\mathbf{x})P(\mathbf{x},t)\} + \frac{1}{2}\sum_{v=1}^{V-1}\sum_{w=1}^{V-1} \frac{\partial^2}{\partial x_v \partial x_w}\{\alpha_{vw}(\mathbf{x})P(\mathbf{x},t)\} + \cdots. \quad (3.4)$$

Here, α_v and α_{vw} are the first and second jump moment of the change in vocabulary that occurs during one iteration of the algorithm, $\delta\mathbf{x}(t) \equiv \mathbf{x}(t+\delta t) - \mathbf{x}(t)$:

$$\alpha_v(\mathbf{x}) = \lim_{\delta t \to 0} \frac{\langle \delta\mathbf{x}_v(t) \rangle}{\delta t}, \quad (3.5)$$

$$\alpha_{vw}(\mathbf{x}) = \lim_{\delta t \to 0} \frac{\langle \delta\mathbf{x}_v(t)\delta\mathbf{x}_w(t) \rangle}{\delta t}. \quad (3.6)$$

In the following, we will go through the calculations of Baxter et al. [1], beginning with the single speaker case due to its simpler form. We will then complete the picture with numerical results obtained from Monte Carlo simulations.

3. Utterance Selection Model of language change

3.1 Single speaker dynamics

When considering only one speaker, the term preceded by H_{ij} in Eq. (3.3) no longer exists, so the update rule has the form

$$\mathbf{x}(t+\delta t) = \frac{\mathbf{x}(t) + (\lambda/T)\mathbf{n}(t)}{1+\lambda}. \tag{3.7}$$

Then the change in vocabulary occurring during one algorithm iteration is

$$\delta \mathbf{x}(t) = \frac{\lambda}{1+\lambda}\left(\frac{\mathbf{n(t)}}{T} - \mathbf{x}(t)\right). \tag{3.8}$$

If the token production is unbiased, then $\mathbf{x}' = \mathbf{x}$. From Eq. (3.2) we obtain $\langle \mathbf{n}(t) \rangle = T\mathbf{x}'$, and therefore the first jump moment of $\delta \mathbf{x}(t)$ vanishes, whereas the second is found to be

$$\langle \delta \mathbf{x}_v(t) \delta \mathbf{x}_w(t) \rangle = \frac{\lambda^2}{(1+\lambda)^2}\frac{1}{T}(x_v \delta_{v,w} - x_v x_w). \tag{3.9}$$

When moving from a discrete model to a continuous one, the time step has to be defined. The model parameters that one can make use of are T or λ. Since taking $\delta t = 1/T$ would make the approximation valid only for a large number T of words uttered per interaction, the parameter λ is a better choice, and therefore:

$$\lambda = (\delta t)^{1/2}. \tag{3.10}$$

The parameters H_{ij} and M_{ij} need to be rescaled too:

$$H_{ij} = h_{ij}(\delta t)^{1/2}, \tag{3.11}$$
$$M_{vw} = m_{vw}(\delta t)^{1/2} \text{ for } v \neq w. \tag{3.12}$$

3.1 Single speaker dynamics

In the biased reproduction case, taking into account the fact that the columns of the matrix M are normalized to 1, we see that all but one element in each column are independent. Choosing this element to be the diagonal one (given by $M_{vv} = 1 - \sum_{w \neq v} M_{wv}$), we can write

$$x'_v - x_v = \sum_w M_{vw} x_w - \sum_w M_{wv} x_v = \sum_{w \neq v}(M_{vw} x_w - M_{wv} x_v). \quad (3.13)$$

Then the first jump moment is

$$c_v(\mathbf{x}) = \sum_{m \neq n}(m_{vw} x_w - m_{wv} x_v). \quad (3.14)$$

From Eq. (3.9) we see that $\langle \delta x_v(t) \delta x_w(t) \rangle$ is of the order δt, so in the limit $\delta t \to 0$ all off-diagonal entries of the matrix M vanish, and hence bias can be ignored. Then the second jump moment is given by

$$a_{vv}(\mathbf{x}) = \frac{1}{T^2}(\langle n_v n_w \rangle - \langle x_v \rangle \langle x_w \rangle). \quad (3.15)$$

The variance of the multinomial distribution, Eq. (3.2), is

$$\langle n_v n_w \rangle - \langle n_v \rangle \langle n_w \rangle = \begin{cases} T x'_v(1 - x'_v), & v = w, \\ -T x'_v x'_w, & v \neq w, \end{cases} \quad (3.16)$$

so Eq. (3.15) turns into

$$a_{vw}(\mathbf{x}, t) = \frac{1}{T}(x_v \delta_{v,w} - x_v x_w). \quad (3.17)$$

Since all higher jump moments are of higher order than δt, they vanish, and therefore the Fokker-Planck equation for the single speaker model is obtained with the aid of the first two moments only:

3. Utterance Selection Model of language change

$$\frac{\partial P(\mathbf{x},t)}{\partial t} = -\sum_{v=1}^{V-1}\frac{\partial}{\partial x_v}\sum_{w\neq v}(m_{vw}x_w - m_{wv}x_v)P(\mathbf{x},t)$$
$$+ \frac{1}{2T}\sum_{v,w}\frac{\partial^2}{\partial x_v \partial x_w}(x_v\delta_{v,w} - x_v x_w)P(\mathbf{x},t). \quad (3.18)$$

What we can already anticipate is that if there is no bias in the token production, eventually all but one variant will die out, so the only possible stationary distribution is one where a variant has frequency 1 and all others are absent. We will now study Eq. (3.18) first in the unbiased, and then in the biased production case.

3.1.1 Unbiased token production

If the speaker utters tokens with probabilities corresponding to the frequencies x_v stored in her "grammar", what remains of Eq. (3.18) is

$$\frac{\partial P(\mathbf{x},t)}{\partial t} = \frac{1}{2T}\sum_{v=1}^{V-1}\sum_{w=1}^{V-1}\frac{\partial^2}{\partial x_v \partial x_w}(x_v\delta_{v,w} - x_v x_w)P(\mathbf{x},t). \quad (3.19)$$

Since T appears as a time scale, in the following it will be set to 1 without loss of generality. The exact solution of Eq. (3.18) can be found through a change of variables [88]:

$$u_i = \frac{x_i}{1 - \sum_{j<i} x_j}. \quad (3.20)$$

Inserting this into Eq. (3.19) results in

$$\frac{\partial P(\mathbf{u},t)}{\partial t} = \frac{1}{2}\sum_{v=1}^{V-1}\frac{\partial^2}{\partial u_v^2}\frac{u_v(1-u_v)}{\prod_{w<v}(1-u_w)}P \equiv \hat{\mathcal{D}}_V(u_1,...,u_{V-1})P. \quad (3.21)$$

3.1 Single speaker dynamics

Separating the time and space variables, for a given initial condition $\mathbf{u}_0$ we can write

$$P(\mathbf{u},t) = \sum_{\lambda_V} C_{\lambda_V}(\mathbf{u}_0) \Phi_{\lambda_V}(\mathbf{u}) e^{-\lambda_V t}, \qquad (3.22)$$

where λ and Φ_{λ_V} are the eigenvalues and eigenfunctions of the operator $\hat{\mathcal{D}}$, and $C_{\lambda_V}(\mathbf{u}_0)$ the expansion coefficients to be determined from the initial condition. Now the variables u can be separated recursively:

$$\hat{\mathcal{D}}_{V+1}(u_1,\ldots,u_V) = \hat{\mathcal{D}}_2(u_1) + \frac{1}{1-u_1} \hat{\mathcal{D}}_V(u_2,\ldots,u_V). \qquad (3.23)$$

Assuming that $\Phi(u_1,\ldots,u_{V-1})$ is an eigenfunction and that λ is an eigenvalue of $\hat{\mathcal{D}}_V(u_1,\ldots,u_{V-1})$, the following ansatz is made for an eigenfunction of $\hat{\mathcal{D}}_{V+1}(u_1,\ldots,u_V)$, with the eigenvalue λ_{V+1} to be determined

$$\Phi_{\lambda_{V+1}}(u_1,\ldots,u_V) = \psi_{\lambda_{V+1},\lambda_V}(u_1) \Phi_{\lambda_V}(u_2,\ldots,u_V). \qquad (3.24)$$

Using this expression in Eq. (3.23) leads to

$$\frac{1}{2}\frac{d^2}{du^2} u(1-u) \psi_{\lambda_{V+1},\lambda_V}(u) = \left(\lambda_{V+1} - \frac{\lambda_V}{1-u}\right) \psi_{\lambda_{V+1},\lambda_V}(u), \qquad (3.25)$$

which has to be solved for the function ψ. If there are only two competing variants (i.e. $V = 2$), only the variable u_1 is independent, and the solution of Eq. (3.25) with $\lambda_1 = 0$ is the eigenfunction $\hat{\mathcal{D}}(u_1)$ with eigenvalue λ_2. Inserting this result into Eq. (3.24) and reiterating, the eigenfunction of the Fokker-Planck Eq. (3.19) is

$$\Phi_{\lambda_{V-1}} = \psi_{\lambda_V,\lambda_{V-1}}(u_1) \psi_{\lambda_{V-1},\lambda_{V-2}}(u_2) \cdots \psi_{\lambda_2,\lambda_1}(u_{V-1}). \qquad (3.26)$$

The eigenvalues of Eq. (3.25) can be found, after an appropriate substitution, by bringing the equation into a standard hypergeometric form with Jacobi polynomials as solution [88]:

$$\lambda_V = \frac{1}{2}L_{V-1}(L_{V-1}+1), \quad L_v = \sum_{w=1}^{v}(l_w+1), \qquad (3.27)$$

where l_w are non-negative integers.

The eigenvalues are positive, which means that the function $P(\mathbf{u},t)$ decays with time. This happens because for unbiased token production a variant with frequency zero can never be uttered again, so eventually all but one variant will go extinct. Due to the conservation of the mean of the distribution, the probability for a particular variant to fixate equals that variant's mean frequency in the initial distribution.

3.1.2 Biased token production

If there is bias at work during token production, the calculations become more difficult, since many more terms have to be taken into account. Under these circumstances, analytical results can be obtained if one imposes the condition that $m_{wv} = m_v$, i.e. the mutation rates depend only on the variant that the others mutate into. In order to calculate the moments of $x_v(t)$, the Fokker-Planck equation is rewritten as a continuity equation: $\partial P/\partial t + \sum_{i,v} \partial J_{iv}/\partial x_{iv} = 0$. Imposing the condition that the current on the boundaries is zero, the equation for the moments reads

$$\frac{d\langle x_v(t)^k\rangle}{dt} = \int d\mathbf{x}\, x_v^k \frac{\partial P(\mathbf{x},t)}{\partial t} = -\sum_w \int d\mathbf{x}\, x_v^k \frac{\partial J_w}{\partial x_w} = k\int d\mathbf{x}\, x_v^{k-1} J_v(\mathbf{x},t), \qquad (3.28)$$

with the probability current

3.1 Single speaker dynamics

$$J_{iv}(X,t) = -\sum_{\substack{w\neq v\\w=1}}^{V}(m_{wv}x_{iv} - m_{vw}x_{iw})P(X,t) \qquad (3.29)$$

$$-\frac{1}{2}\sum_{w=1}^{V-1}\frac{\partial}{\partial x_{iv}}(x_{iv}\delta_{v,w} - x_{iv}x_{iw})P(X,t). \qquad (3.30)$$

The equation for the first moment is then

$$\frac{d\langle x_v(t)\rangle}{dt} = -\sum_{w\neq v}(m_w\langle x_v\rangle - m_v\langle x_w\rangle) = \left(-\sum_{w\neq v}m_w\right)\langle x_v\rangle + m_v(1-\langle x_v\rangle)$$
$$= m_v - R\langle x_v\rangle, \qquad (3.31)$$

where $R = \sum_{v=1}^{V} m_v$. The solution of Eq. (3.31) is

$$\langle x_v(t)\rangle = \frac{m_v}{R} + \left(x_{v,0} - \frac{m_v}{R}\right)e^{-Rt}. \qquad (3.32)$$

The full time-dependent solution of the Fokker-Planck Eq. (3.19) can be found by the same change of variables as in the unbiased case, Eq (3.20). One has to replace

$$\frac{1}{2}\frac{\partial}{\partial u_v}u_v(1-u_v) \rightarrow \frac{1}{2T}\frac{\partial}{\partial u_v}u_v(1-u_v) + (R_v u_v - m_v), \qquad (3.33)$$

where $R_v = \sum_{w=v}^{V} m_w$ in Eq. (3.21), and then the eigenvalues become

$$\lambda_V = \frac{1}{2T}L'_{\gamma-1}(2TR + L'_{M-1} - 1), \quad L'_v = \sum_{w=1}^{v} l_w, \qquad (3.34)$$

with l_w non-negative integers as before. Here there is a zero eigenvalue when $l_w = 0\ \forall w$. The eigenfunction corresponding to it is the stationary state

3. Utterance Selection Model of language change

$$P^*(\mathbf{x}) = \Gamma(2R)\prod_{v=1}^{V}\frac{x_v^{2Tm_v-1}}{\Gamma(2m_v)}, \qquad (3.35)$$

which for $V = 2$ is a β distribution. This is convex when m_1 and m_2 are less than $1/2$, and concave otherwise. If $m_1 \neq m_2$, the distribution is asymmetric.

3.2 Dynamics of N speakers

Returning to the initial model with N interacting speakers, the calculation goes on similar lines as for one speaker.

From the update rule Eq. (3.3),

$$\delta x_{iv} = \frac{\lambda}{1+\lambda(1+H_{ij})}\left[\frac{n_{iv}}{T} - x_{iv} + H_{ij}\left(\frac{n_{jv}}{T} - x_{iv}\right)\right]. \qquad (3.36)$$

Given that $\langle n_{iv}\rangle = Tx'_{iv}$,

$$\begin{aligned}\langle \delta x_{iv}\rangle &= \frac{\lambda}{1+\lambda(1+H_{ij})}[x'_{iv} - x_{iv} + H_{ij}(x'_{jv} - x_{iv})]\\ &= \lambda\left(\sum_{w\neq v}(M_{vw}x_{iw} - M_{wv}x_{iv}) + H_{ij}(x_{jv} - x_{iv})\right) + O(\lambda HM, \lambda^2 H, \lambda^2 M).\end{aligned} \qquad (3.37)$$

For the N-speaker model, the variance of the multinomial distribution is

$$\langle n_v n_w\rangle - \langle n_v\rangle\langle n_w\rangle = \begin{cases} Tx'_{iv}(1-x'_{iv}), & v=w,\ i=j \\ -Tx'_{iv}x'_{iw}, & v\neq w,\ i=j \\ 0, & i\neq j.\end{cases} \qquad (3.38)$$

With this the second moment is found:

3.2 Dynamics of N speakers

$$\langle \delta x_{iv} \delta x_{jw} \rangle = \frac{\lambda^2}{T}(x_{iv}\delta_{v,w} - x_{iv}x_{iw}) + O(\lambda^2 H, \lambda^2 M, \lambda^3) \rangle \quad (3.39)$$

for $i = j$ and $\langle \delta x_{iv} \delta x_{jw} \rangle = 0$ otherwise.

Since the Fokker-Planck equation consists of a deterministic and a stochastic term, both $\langle \delta x_{iv} \rangle$ and $\langle \delta x_{iv} \delta x_{iw} \rangle$ have to be of order δt in the limit $\delta t \to 0$. This is ensured by the rescaling of variables Eqs. (3.10) and (3.12). Again, as in the single speaker case, all higher order terms vanish, so by inserting the calculated jump moments into Eq. (3.4), taking into consideration the interaction probabilities G_{ij} and averaging over all pairs of speakers, the general Fokker-Planck equation reads:

$$\frac{\partial P(X,t)}{\partial t} = \sum_i G_i [\hat{\mathcal{L}}_i^{(bias)} + \hat{\mathcal{L}}_i^{(rep)}] P(X,t) + \sum_{\langle ij \rangle} G_{ij} \hat{\mathcal{L}}_{ij}^{(int)} P(X,t), \quad (3.40)$$

where $G_i = \sum_{j \neq i} G_{ij}$ is the probability that speaker i takes part in an interaction with any other speaker.

The term

$$\hat{\mathcal{L}}_i^{(bias)} = \sum_{v=1}^{V-1} \frac{\partial}{\partial x_{iv}} \sum_{\substack{w=1 \\ w \neq v}}^{V} (m_{wv} x_{iv} - m_{vw} x_{iv}) \quad (3.41)$$

represents the effect of bias in the reproduction of variants.

Due to the fact that the number of utterances produced in an interaction is finite, stochasticity is present in the Fokker-Planck equation through the term

$$\hat{\mathcal{L}}_i^{(rep)} = \frac{1}{2T} \sum_{v=1}^{V-1} \sum_{w=1}^{V-1} \frac{\partial^2}{\partial x_{iv} \partial x_{iw}} (x_{iv} \delta_{v,w} - x_{iv} x_{iw}). \quad (3.42)$$

3. Utterance Selection Model of language change

The last term,

$$\hat{\mathcal{L}}_{ij}^{(int)} = \sum_{v=1}^{V-1} \left(h_{ij} \frac{\partial}{\partial x_{iv}} - h_{ji} \frac{\partial}{\partial x_{jv}} \right)(x_{iv} - x_{jv}), \qquad (3.43)$$

arises from the influence of the interlocutor's utterances on the speaker's vocabulary.

In order to be able to gain some insight by means of analytical calculations, the number of parameters is reduced by setting all H_{ij} equal to a constant h, the number of variants is restricted to two, and all weights on the links connecting speakers are equal:

$$G_{ij} \equiv G = \frac{1}{2N(N-1)} \quad \forall i,j. \qquad (3.44)$$

The Fokker-Planck Eq. (3.40) becomes

$$\begin{aligned}
\frac{\partial}{\partial t} P &= (N-1)G \sum_i \left(\frac{\partial}{\partial x_i}(Rx_i - m_1) + \frac{1}{2T} \frac{\partial^2}{\partial x_i^2} x_i(1 - x_i) \right. \\
&\quad \left. + h \frac{\partial}{\partial x_i}\left(x_i - \frac{1}{N-1} \sum_{j \neq i} x_j \right) \right) P \\
&= (N-1)G \sum_i \left(\frac{\partial}{\partial x_i}(Rx_i - m_1) + \frac{1}{2T} \frac{\partial^2}{\partial x_i^2} x_i(1 - x_i) \right. \\
&\quad \left. + \frac{N}{N-1} h \frac{\partial}{\partial x_i}(x_i - x) \right) P, \qquad (3.45)
\end{aligned}$$

where x represents the average frequency of the first variant throughout the population, $x \equiv \sum_i x_i/N$, $m_1 \equiv m_{12}$ and $R = m_1 + m_2 = m_{12} + m_{21}$.

In the same way as above, one can find equations for the moments x_i:

3.2 Dynamics of N speakers

$$\frac{d}{dt}\langle x_i \rangle = -(N-1)G\left((R+h)\langle x_i \rangle - m_1 - \frac{h}{N-1}\sum_{j \neq i}\langle x_i \rangle\right). \tag{3.46}$$

Writing the sum over j as $N\langle x \rangle - \langle x_i \rangle$, with $\langle x \rangle = 1/N \sum_i \langle x_i \rangle$ and summing Eq. (3.46) over all speakers, we obtain

$$\frac{d}{dt}\langle x \rangle = -G(N-1)(R\langle x \rangle - m_1). \tag{3.47}$$

Then, subtracting this from Eq. (3.46), we arrive at

$$\frac{d}{dt}\langle x_i - x \rangle = -G[(N-1)(R+Nh)\langle x_i - x \rangle. \tag{3.48}$$

These decoupled equations have the following solution:

$$\langle x_i(t) \rangle = \frac{m_1}{R} + \left[\left(x_0 - \frac{m_1}{R}\right) + (x_{i,0} - x_0)e^{-ht/2(N-1)}\right]e^{-Rt/2N}, \tag{3.49}$$

$$\langle x(t) \rangle = \frac{m_1}{R} + \left(x_0 - \frac{m_1}{R}\right)e^{-Rt/2N}, \tag{3.50}$$

with $x_0 = x(0) = \frac{1}{N}\sum_i x_{i,0}$.

In the unbiased case, this gives

$$\langle x_i(t) \rangle x_0 + (x_{i,0} - x_0)e^{-ht/2(N-1)} \tag{3.51}$$

and

$$\langle x(t) \rangle = x_0, \tag{3.52}$$

so the average frequency of a variant throughout the population is conserved. Eventually one of the variants will die out, leaving the speech community in

3. Utterance Selection Model of language change

consensus over the other one. This time can be calculated by using the method of the coalescent. The time until a particular variant (e.g. the second) dies out is found to be

$$\tau_2[X(0)] = \frac{1-x_0}{x_0}\left(\frac{N(N-1)}{2h}F[X(0)] - TN^2\ln(1-x_0)\right). \tag{3.53}$$

The function F depends on the initial configuration of the system, so if all speakers start with the same initial proportion of a variant, $x_i(0) = x_0\ \forall i$,

$$F[X(0)] = \sum_{m=1}^{N-1} \frac{x_0^m}{m} - \frac{x_0}{N}\frac{1-x_0^{N-1}}{1-x_0}, \tag{3.54}$$

and if $M = Nx_0$ of the speakers start with $x_i = 1$ and $N - M$ start with $x_i = 0$,

$$F[X(0)] = \sum_{m=1}^{M} \frac{\binom{M}{m}}{\binom{N}{m}}\frac{1}{m}. \tag{3.55}$$

For large N, the values of F can be approximated by

$$F[X(0)] \sim -\ln(1-x_0), \tag{3.56}$$

which then leads to

$$\tau_2 \sim -\frac{1-x_0}{x_0}\ln(1-x_0)\left(\frac{N(N-1)}{2h} + TN^2\right). \tag{3.57}$$

The fixation time of any of the variants is then found by taking a weighted average of the time for each variant:

$$t_c = x_0\tau_2 + (1-x_0)\tau_1 \sim -[(1-x_0)\ln(1-x_0) + x_0\ln x_0]\left(\frac{N(N-1)}{2h} + TN^2\right). \tag{3.58}$$

3.3 Monte Carlo simulations

We will now explore the Utterance Selection Model with complementary tools, namely numerical simulations. For this, we will restrict the number of variants to $V = 2$, as in the last part of the previous section. Since the numerics seldom provides round numbers, we define consensus to be reached when the average frequency of one of the two variants is within an error range to either 0 or 1. The error value we use is 10^{-8}, which means that if $0.9999999 < x \leq 1$, then for us $x = 1$.

3.3.1 Unbiased production

In the analytical approach presented in the previous section, by taking the continuous-time limit, one of the parameters of the model, λ, was fixed in order to define the time step. The parameter λ gives the change pressure exerted by an interaction on the vocabulary of the speaker. For very small λ, the new frequency values differ only slightly from the old ones, and the time to consensus is very large. From fitting the numerical data in Figure 3.4, for small λ we find that the average time to consensus is proportional to $1/\lambda^2$. Since a speaker's vocabulary does not undergo dramatic changes in the course of an interaction, the value of λ should be small, but still large enough to enable us an adequate number of simulation runs. Baxter et al. [1] indicate that for $\lambda \sim 10^{-3}$, the analytical findings and the simulation results show a good match. For most of our simulations we will use $\lambda = 0.01$, which provides results that agree with previous findings. At the other end of the spectrum, for $\lambda \gg 1$, the dynamics becomes independent of λ (Figure 3.4) and the old frequencies only play a role through the uttered tokens, so the update rule turns into

$$\mathbf{x}_i(t + \delta t) \approx \frac{\frac{1}{T}\left(\mathbf{n}_i(t) + H_{ij}\mathbf{n}_j(t)\right)}{1 + H_{ij}}. \tag{3.59}$$

3. Utterance Selection Model of language change

If T is large, $\mathbf{n}_i/T \approx \mathbf{x}_i$, therefore the new frequencies of speaker i will be composed of her old frequencies and those of her interlocutor (with weight H_{ij}). If T is small, meaning that the utterances are short, then the preferred variants of the speaker and of her interlocutor have good chances of being the only ones to contribute to the new frequencies.

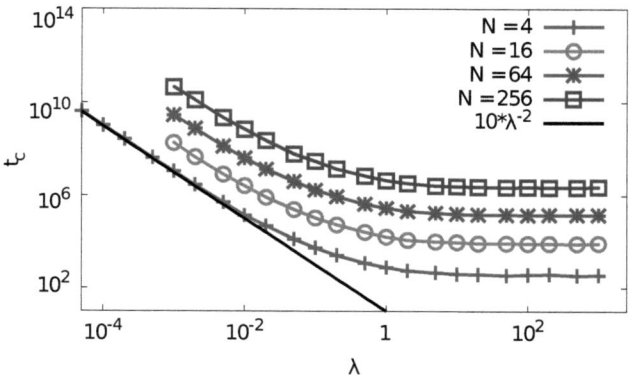

Figure 3.4: The average time to consensus in a group of N speakers as a function of the change pressure λ. If λ is small, $t_c \propto 1/\lambda^2$, whereas for large values of the parameter the dynamics becomes independent of λ. Other parameter values are $T = 1$, $h = 0.01$, the data points are averages over 1000 simulation runs.

We will now explore the role of another parameter of the Utterance Selection Model, namely H_{ij}. It stands for the weight given by a speaker to her interlocutor's utterances relative to her own. As in the analytic approach, first we will set all H_{ij} equal to a constant h. From Eq. (3.58) we see that the time until a variant fixates in the speech community is proportional to $1/h$. This approximation holds for small values of h, when the interlocutor has a lower status than the speaker herself. Increasing h, we see that the curves display a plateau around $h = 1$. Here, the utterances of the two speakers have the same weight, and the update rule Eq. (3.3) becomes

3.3 Monte Carlo simulations

$$\mathbf{x}_i(t+\delta t) = \frac{\mathbf{x}_i(t) + (\lambda/T)[\mathbf{n}_i(t) + \mathbf{n}_j(t)]}{1+2\lambda}. \tag{3.60}$$

Moving towards higher values of h, we discover that when $\lambda \cdot h \approx 1$, the average time to consensus reaches a minimum (Figure 3.5). Under these circumstances, Eq. (3.3) has the form

$$\mathbf{x}_i(t+\delta t) = \mathbf{x}_i(t) + \mathbf{n}_j(t)/T. \tag{3.61}$$

This means that the most favorable conditions for consensus to be reached are when the utterances of the interlocutor have the same weight as the speaker's old set of frequencies, and the vocabularies of the speakers are aligned as much as allowed in one interaction. Letting h take higher values, the time to consensus increases again. Now the old frequencies are no longer important, the update rule now turning into

$$\mathbf{x}_i(t+\delta t) \approx \mathbf{n}_j(t)/T. \tag{3.62}$$

This is because in this regime, the speaker adopts the utterances of her interlocutor as her new set of frequencies, with her old vocabulary being neglected. As the interaction is symmetric, the same happens for the interlocutor. Since they "swap" frequencies and by this adjust their vocabularies in different directions, reaching an agreement becomes an increasingly difficult task, for $h \to \infty$ even impossible. In this regime, the time to consensus is proportional to h. The laws governing the small-h and large-h regimes are symmetric, reflecting how a certain amount of influence ascribed to a speaker can determine a vocabulary, be it her own ($h \ll 1$) or the one of her interaction partner($h \gg 1$).

Since in a real society not all speakers have the same social status, some of them might have more influence than others on the community's vocabulary. A question that arises at this point is how the time to consensus changes if the values of H_{ij} are not equal, but drawn from a distribution. We will first

3. Utterance Selection Model of language change

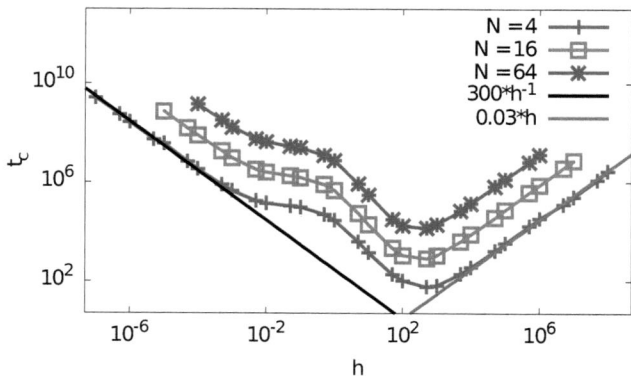

Figure 3.5: The average time to consensus for N speakers as a function of $H_{ij} = h$. In the case of small h, $t_c \propto 1/h$. When $\lambda \cdot h \approx 1$, the average time to consensus reaches a minimal value. If h is large, t_c increases proportional to h. Other parameter values: $\lambda = 0.01$, $T = 1$, data averaged over 1000 simulation runs.

consider a uniform distribution around a particular value, which in this case we choose to be $H_{ij} = h = 0.01 \quad \forall i, j$. Although the differences between the curves are not very pronounced, the simulation results suggest that a broader distribution favors faster consensus (Figure 3.6a).

In a different scenario, the values H_{ij} are drawn from a Gaussian distribution. For a narrow distribution, we recover the case of constant H_{ij}'s. The numerical results show that with broadening distribution the consensus time becomes shorter (Figure 3.6b).

Finally, we draw H_{ij} from a Pareto distribution [89]. If the parameter α is between 1 and 2, one can easily generate numbers with a predefined expectation value. What we see here is that the flatter the distribution is (i.e. the lower α), the faster consensus is reached (Figure 3.6c). This means that speakers with intermediate values of H_{ij} have a considerable weight in the process of agreeing on a particular variant.

From these three cases we learn that the more heterogenous a society is, the

3.3 Monte Carlo simulations

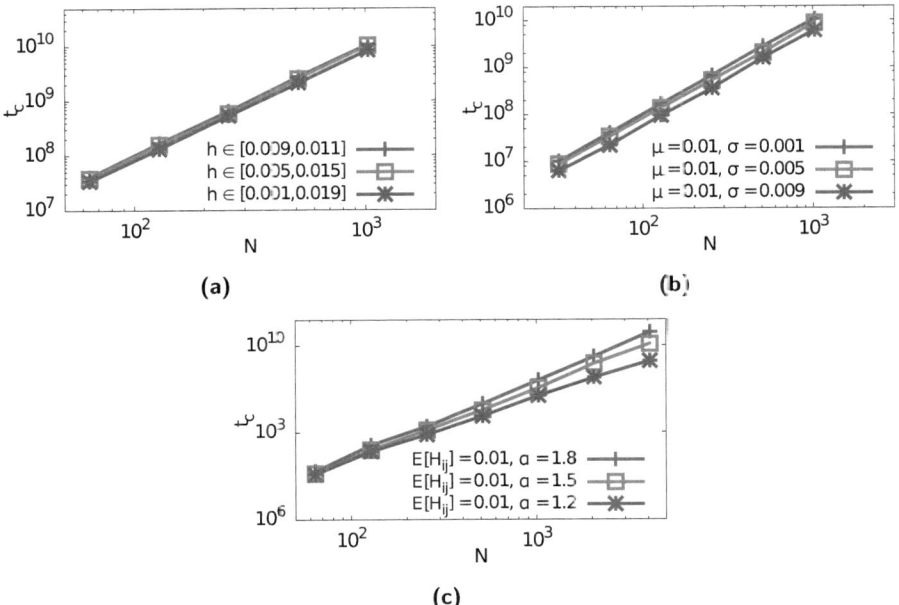

Figure 3.6: H_{ij}'s drawn from various distributions. (a) Uniform distribution around $h = 0.01$ (b) Gaussian distribution with mean $h = 0.01$ (c) Pareto distribution with expectation value $E[H_{ij}] = 0.01$. In all three cases, a broader distribution favors a shorter time to consensus.

faster it will find a linguistic variant accepted by all its members. As for the various distributions we considered, although it is difficult to compare them with each other, the Pareto-distributed values of H_{ij} have induced, for large systems, a time to consensus almost one order of magnitude lower than the ones drawn from the uniform and the Gaussian distributions respectively.

The parameter T, i.e. the length of the strings of words uttered by the speakers, contributes to the stochasticity of the process, since a large T returns the average frequencies of the speaker, whereas a small T usually promotes the more frequent variant. From simulations we recover the result of Baxter et al,

3. Utterance Selection Model of language change

namely that the dependence of the time to consensus on T is linear (Figure 3.7).

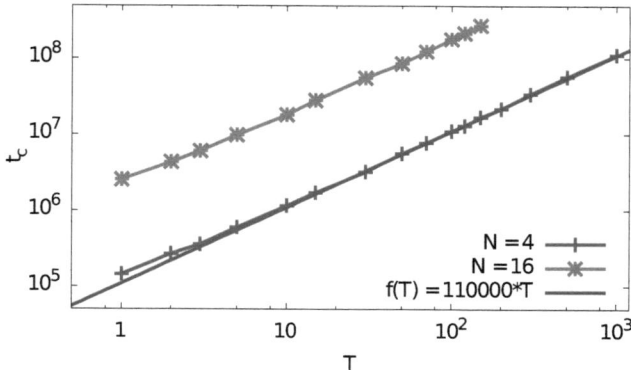

Figure 3.7: Average time to consensus as a function of the length of utterances T for systems of $N = 4$ and 16 speakers. The simulation data coincides with the analytic linear dependence. Data points are averages over 1000 runs, the other parameters are $h = 0.01$ and $\lambda = 0.01$.

3.3.2 The role of network topology

Finally, we ask how the topology of the speaker network influences the time to consensus. For this, we run simulations on a well-mixed network, on a square lattice with periodic boundaries, and on an uncorrelated scale-free random network (the details on the algorithm for generating these networks are presented in Appendix A). From Figure 3.8 we learn that on the well-mixed networks, consensus is the fastest. On a square lattice it is still faster than on the random network. It appears however that on all three networks, the dependence on the system size is squared. These results for finite sized networks are in good agreement with those of Baxter [90], who shows that decreasing the path length between speakers reduces the time for the fixation of a variant. On the

3.3 Monte Carlo simulations

other hand, Baxter et al. [91] show that in the thermodynamic limit the time to consensus is network-independent.

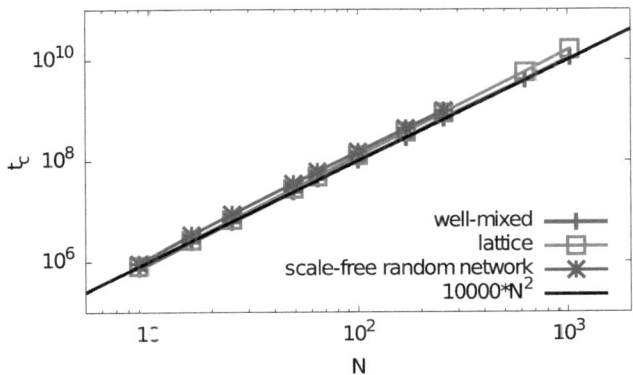

Figure 3.8: Time to consensus as a function of the system size for three different types of networks: fully connected, square lattice and scale-free random network with $\gamma = 3$. Data points are averages over 1000 runs, the other parameters are $T = 1$, $\lambda = 0.01$, $h = 0.01$.

3.3.3 Diffusion on a square lattice

Another question we will answer with the aid of stochastic simulations is how the time to consensus behaves if the speakers are not fixed to their lattice sites but can diffuse around the system. To simulate this, we will use the Gillespie algorithm, presented in more detail in Appendix B. A diffusion step consists in choosing a speaker at random and swapping her position with a randomly chosen neighbor. This way, speakers move through the system at low speed. Numerical results show that with increasing diffusion coefficient, consensus is reached slightly faster, but the effect is small. The data shows a considerable amount of noise, so that we cannot establish the exact dependence of the time to consensus on the diffusion coefficient d. One can see this setting as an

3. Utterance Selection Model of language change

intermediate state between a stationary group on a two-dimensional lattice and a well-mixed group.

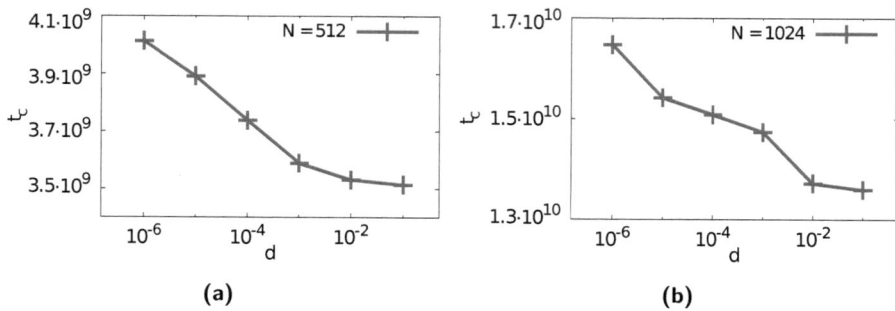

Figure 3.9: Average time to consensus as a function of the diffusion coefficient for (a) $N = 512$ and (b) $N = 1024$. Other parameter values: $T = 1$, $\lambda = 0.01$, $h = 0.01$. The data points are averages over 1000 simulation runs.

3.3.4 Biased production

For biased variant production we can also obtain some insights from numerical simulations. In the following, we will assume a symmetric mutation matrix. Baxter et al. [1] find the single speaker stationary distribution for the two-variant model to be a β distribution:

$$P^*(x) \approx \frac{\Gamma(2\alpha)}{\Gamma(\alpha)^2}[x_i(1-x_i)]^{\alpha-1} \quad \text{with} \quad \alpha = 2T\frac{m}{\lambda}\left(\frac{(N-1)m + Nh}{(N-1)m + h}\right), \quad (3.63)$$

with the parameters m and h being the ones used in simulations, representing the original M_{ij} and H_{ij}, and not the rescaled ones used by Baxter et al. in the analytical calculations. This distribution is convex (has a peak near each boundary) if bias is small, and becomes centrally peaked for large m (Figure 3.10).

3.3 Monte Carlo simulations

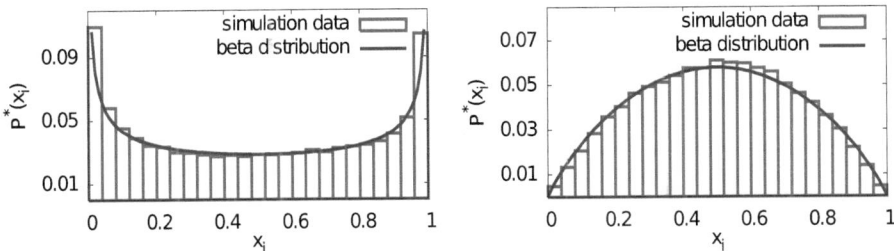

Figure 3.10: Beta distribution for a system of $N = 10$ speakers with $m = 0.0002$ (left) and $m = 0.0009$ (right). Other parameters: $\lambda = 0.001$, $h = 0.0002$, $T = 1$.

There is a critical mutation rate m_c, where the distribution switches from convex to concave, and this is the point where $\alpha = 1$. The simulation results show that the analytically found curve for the critical mutation rates holds only for $h < 1$ (Figure 3.11). Having previously assumed that T could be set to 1 w.l.o.g., we initially ascribed this deviation of the simulations from the theory to the particular case $T = 1$. It appears that for $T = 10$ the simulations still show results different from theory (Figure 3.12).

The differences between the beta distribution and the simulation data become visible around $h = 1$, i.e. in the parameter region where the utterances of the two speakers have equal weight, and amplify from there on. It would be interesting to know whether they decrease again after the point where $\lambda \cdot h \approx 1$, but in our simulations the error bars are too large to allow such observations. One possibility for the appearance of this effect is the asymmetry of the update rule Eq. (3.3), where the old frequencies of the speaker play an important role in slowing down the changes occurring due to other speakers as long as h is small enough, i.e. the influence of the interlocutor is not very strong. Once this barrier is crossed, the update rule becomes Eq. (3.61), and the first term becomes continually weaker. For very large h, the single-speaker as well as the average frequency distribution become uniform, meaning that the speakers

3. Utterance Selection Model of language change

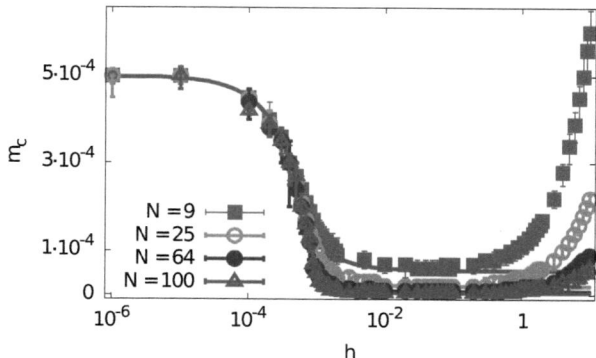

Figure 3.11: Beta distribution (lines) and simulation data (points with error bars) for $T = 1, \lambda = 0.001$. They coincide up to $h \approx 1$, after which, with increasing influence of the interlocutor, the curve described by the data points shows an increasing critical mutation rate. For $h > 1/\lambda$ the error bars are too large to continue observations in this parameter range.

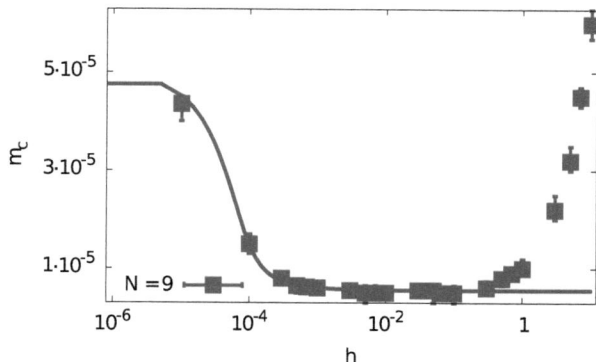

Figure 3.12: Beta distribution (lines) and simulation data (points with error bars) $\lambda = 0.001$, T=10. Here we have the same behavior as for $T = 1$, so this effect is not an artifact of the case of only one token uttered by each speaker during an interactions.

explore the whole frequency spectrum as they continue to interact.

Returning to the other end of the spectrum, small values of h mean that each

3.3 Monte Carlo simulations

speaker will mainly influence herself. Weak bias allows speakers' frequencies to stay at the ends of the interval. The average frequency displays multiple peaks, representing a particular fraction of the speakers with the frequency at one end of the interval (Figure 3.13).

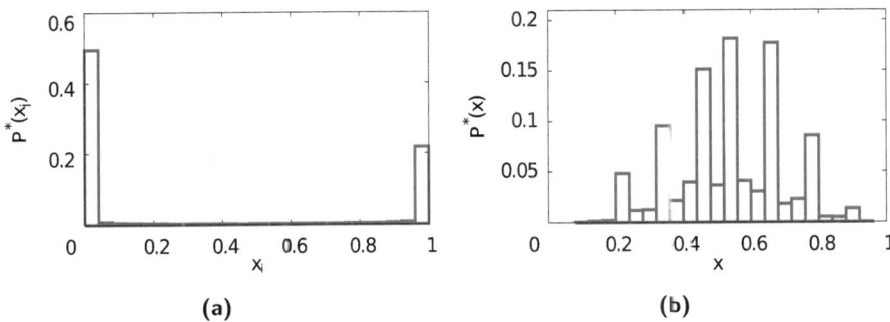

Figure 3.13: (a) Frequency distribution of speaker 1 and (b) average frequency distribution for $h = 0.000001$ and $m = 0.00001$. In (a) the distribution is peaked at the boundaries, whereas the distribution (b) displays multiple peaks representing fractions of speakers with frequencies at one end of the interval. Other parameters: $N = 9$, $\lambda = 0.001$, $T = 1$.

Increasing h, the speakers' influence on each other amplifies, but the individual frequency distribution is still convex. Due to the fact that more and more individuals start favoring the same variant, the average frequency distribution will also become double-peaked at the boundaries (Figure 3.14).

If bias is a little stronger, but still below the critical value, speakers begin switching from one variant to the other, so that the single speaker distribution is still convex, but the average frequency distribution now adopts a concave shape (Figure 3.15).

Very large h implies such a high influence of the conversation partner, that speakers "forget" their old vocabulary and start using the variants uttered by their interlocutors. This becomes reflected in a single speaker as well as an average frequency distribution where only the ends of the interval are populated

3. Utterance Selection Model of language change

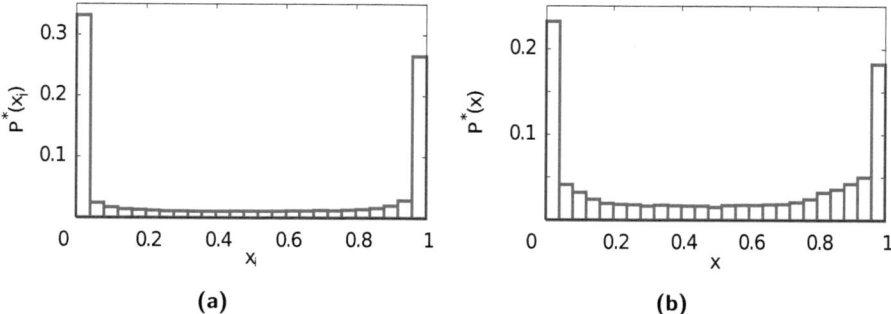

Figure 3.14: (a) Frequency distribution of speaker 1 and (b) average frequency distribution for $h = 0.0005$ and $m = 0.00001$. In (a), the distribution is still convex and (b) has a similar shape due to the fact that the speakers favor the same variant. Other parameters are: $N = 9$, $\lambda = 0.001$, $T = 1$.

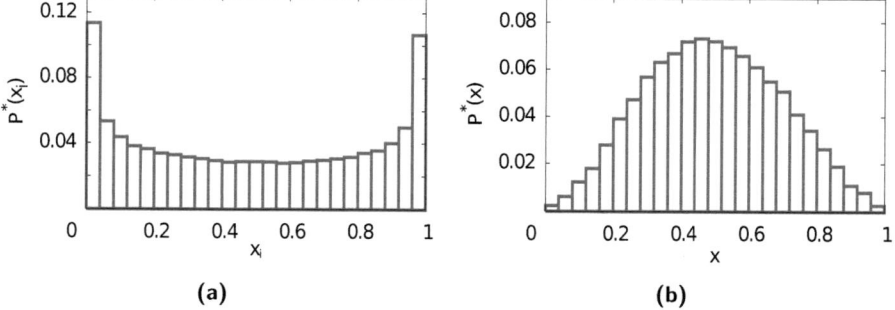

Figure 3.15: (a) Frequency distribution of speaker 1 and (b) average frequency distribution for $h = 0.0005$ and $m = 0.0001$. Since $m < m_c$, the single speaker distribution is still double-peaked at the ends of the interval, but since the variants that each speaker uses differ, the average frequency distribution becomes centrally peaked. Other parameters: $N = 9$, $\lambda = 0.001$, $T = 1$.

(Figure 3.16), since the change after only one interaction is so dramatic, that intermediate values can hardly be reached any more.

If bias is high (above the critical value m_c), speakers will use both variants either at the same time or switching from one to the other, hence displaying a

3.3 Monte Carlo simulations

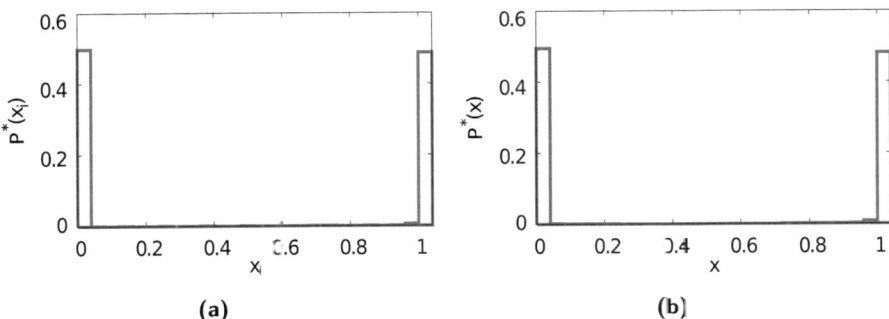

Figure 3.16: (a) Frequency distribution of speaker 1 and (b) average frequency distribution for $h = 1000$ and $m = 0.0001$. For very large h, the speakers set their new frequencies according to the utterances of their interlocutor. Since they only "swap" frequencies but do not tend to align their vocabularies, these do not change very much. The other parameters are: $N = 9$, $\lambda = 0.001$, $T = 1$.

uniform single speaker frequency distribution. The average distribution will be centrally peaked, and the value of h only changes the width of the distribution slightly, this becoming narrower with increasing m (Figures 3.17 and 3.18).

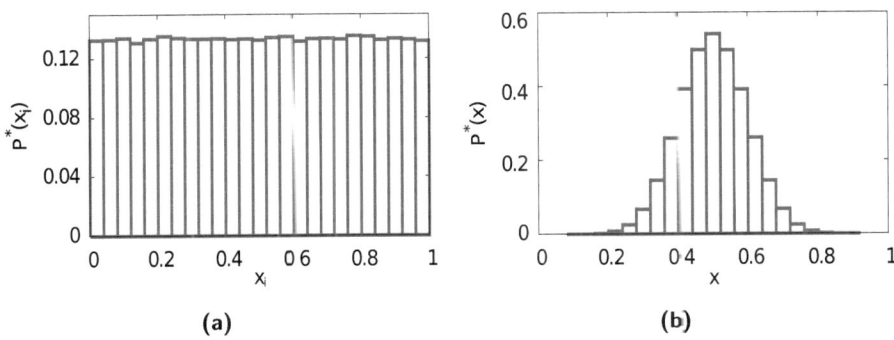

Figure 3.17: (a) Frequency distribution of speaker 1 and (b) average frequency distribution for $h = 1000$, $m = 0.5$. In this case bias is so high that speakers change between variants very often, and thus the single speaker distribution becomes uniform. The average distribution is centrally peaked. Other parameters are: $N = 9$, $\lambda = 0.001$, $T = 1$.

3. Utterance Selection Model of language change

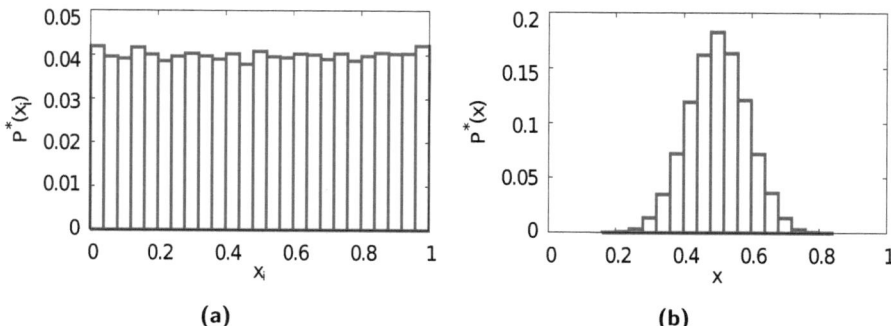

Figure 3.18: (a) Frequency distribution of speaker 1 and (b) average frequency distribution for $h = 1000$, $m = 0.8$. The difference with respect to Figure 3.17 is that for higher bias, the average frequency distribution becomes narrower.

3.4 Utterance Selection Model on a mobile phone network

Wishing to test the Utterance Selection Model on a real world network after having studied its behavior on artificially controlled ones, we will now explore the dynamics on a network of mobile phone users. The available data comprises the phone calls that were made in the network of a European provider of mobile phone services in one month. It contains information about a connected component consisting of 1.044.397 users and 13.983.433 phone calls among these users [92], as well as the duration of these calls. The undirected network has an average degree $\langle k \rangle = 4.24$. On a network this big consensus would take more time than we are able to simulate, so in this case we are not looking for the time to consensus. Instead, we want to find out whether there is a quasi-stationary frequency distribution, and if the answer to this question is affirmative, what characterizes it.

3.4 Utterance Selection Model on a mobile phone network

3.4.1 Quasi-stationary frequency distribution

In order to be able to tell whether a frequency distribution is convex or concave without plotting every data set, we introduce the parameter $r(t) \in [0, 1]$ which gives the fraction of speakers with frequencies in the interval $[0.25, 0.75]$ at a given time t. A large r suggests that the majority of the speakers have frequencies in the center of the interval $[0, 1]$, in other words, they are using both variants regularly. We study the time evolution of this parameter, which tells us whether a quasi-stationary distribution is reached if it has the same value from a particular moment in time on.

The simulation results for different initial conditions ($r(0) = 0$ if speakers use one preferred variant, $r(0) = 0.5$ if the frequencies are spread over the whole spectrum meaning there are "monolingual" as well as "bilingual" speakers, and $r(0) = 1$ if all speakers use both variants. The initial frequencies are spread uniformly in the respective intervals) presented in Figure 3.19 show that the convergence time to the quasi-stationary distribution depends on h. The left plot of Figure 3.19 documents the fact that for small h (speakers influencing each other very little), after 1000 loops[1] the frequency distribution is not yet the same for the three types of initial conditions. For $h = 10$, the quasi-stationary distribution is reached after about 30 loops (right plot), and for $h = 100$ after 1 loop there is already no difference between the curves for different initial conditions. We define a convergence criterium, i.e. a t_{conv} for which the two $r(t)$ corresponding to the initial conditions $r(0) = 0$ and $r(0) = 1$ differ by less that 10%.

Plotting the convergence time as a function of h, we find that for $h > 0.05$, t_{conv} is proportional to $1/h$. For lower h, t_{conv} becomes increasingly independent of this parameter (Figure 3.20).

Having established a convention on when a quasi-stationary distribution is

---

[1] A loop is one passage through the whole phone call database, after which we go back to the beginning of the database.

3. Utterance Selection Model of language change

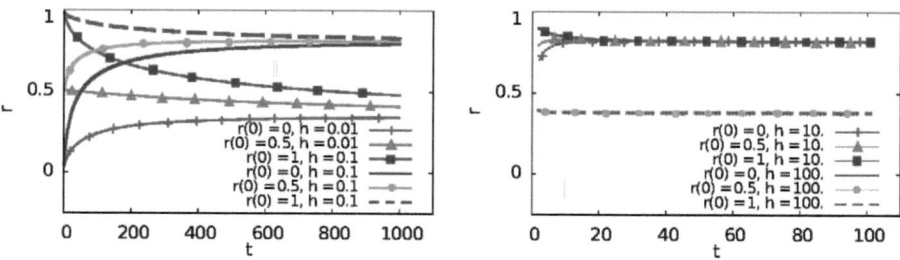

Figure 3.19: Time to convergence on quasi-stationary distribution, given in terms of the number of database loops, for fixed $\lambda = 0.01$ and $T = 1$, four values of h and different initial conditions. Time is given here in number of database loops

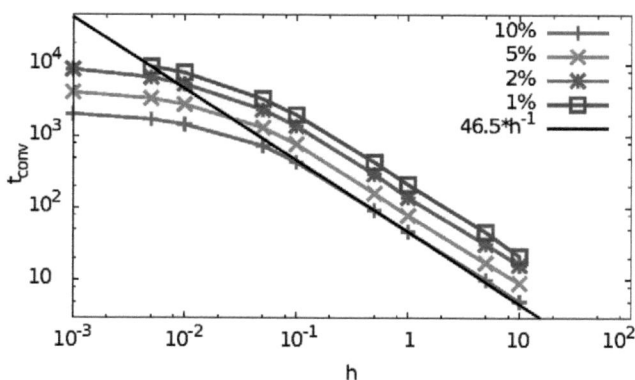

Figure 3.20: Dependence of the convergence time on h for different convergence criteria: the curves $r(0) = 0$ for and $r(0) = 1$ have to differ by less than 1, 2, 5 or 10 percent. Since the qualitative behavior of the four curves is the same, from now on we will use the 10% criterium, since it gives good results while at the same time allowing for shorter simulation times. For $h > 0.05$, t_{conv} is proportional to $1/h$ (here the time is given in number of database loops). Other parameter values: $\lambda = 0.01$, $T = 1$.

reached, we now want to find out what shape this distribution has. A first step towards achieving this goal consists in studying $r(t > t_{conv})$, i.e. the fraction of speakers whose quasi-stationary frequencies are in the interval $[0.25, 0.75]$.

3.4 Utterance Selection Model on a mobile phone network

Plotting this as a function of h for various values of λ (Figure 3.21), we find that increasing h, r first increases and subsequently decreases, for some values of λ going from $r < 0.5$ to $r > 0.5$ and back to lower values. It thus appears that going from low to high values of h, the distribution changes from convex to concave and back to convex.

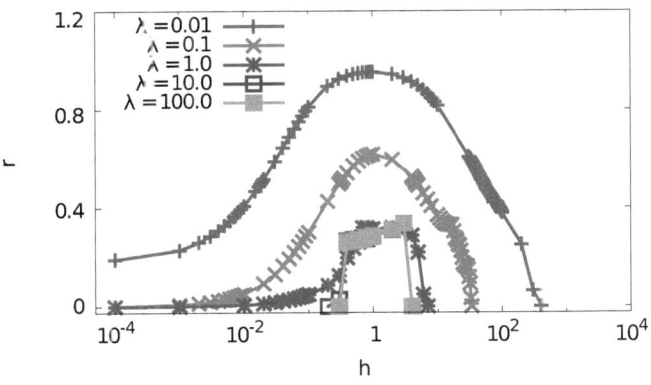

Figure 3.21: Dependence of r on h for various λ after 1000 database loops. The initial condition is r(0)=0.5, for this value the quasi-stationary distribution is reached faster than for r(0)=0 or r(0)=1.

The results of studying the distributions in a broad parameter range are summed up in Figure 3.22. The left curve marks the transition from a convex distribution to a concave one, for $\lambda < 0.2$. The right curve, $\lambda = 1/h$, is delimiting the parameter region that leads to a concave distribution ($\lambda < 1/h$) from the one where the distribution shows multiple peaks ($\lambda > 1/h$), which cannot be described as convex, although $r < 0.5$. To better understand this picture, we study the distributions for two fixed values of λ and several h (in the parameter region marked by the two black horizontal lines in Figure 3.22) in Figure 3.23, and then for two fixed values of h and several λ (marked in Figure 3.22 by the two vertical lines) in Figure 3.24.

In Figure 3.23a, we notice that for $h = 0.001$ the distribution is convex.

3. Utterance Selection Model of language change

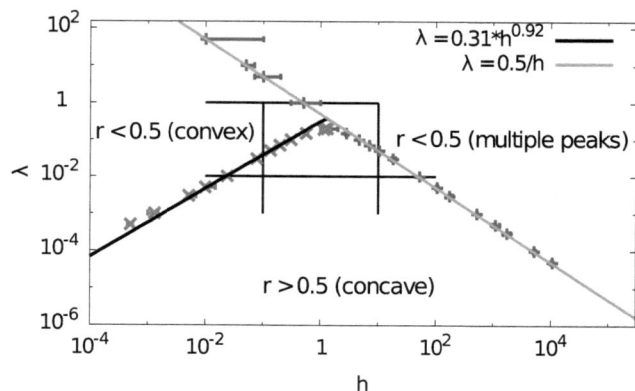

Figure 3.22: (h,λ) diagram. The left curve ($\lambda = h$) delimits the convex/concave distribution regimes. The right curve ($\lambda = 0.5/h$) stands for the transition from a concave to a distribution with multiple peaks. If $\lambda > 0.2$, the transition from convex to multiple peak distribution when increasing h is direct, without passing through the concave distribution regime. The horizontal black lines indicate the range of the parameter h plotted in Figure 3.23, and the vertical ones that of the parameter λ in Figure 3.24.

Then for h equal to 1, the distribution becomes centrally peaked. Increasing h further, the central peak splits into two and then several other peaks that move towards the boundaries of the interval, symmetric with respect to its center. This is because, the larger h is, the more drastic the vocabulary change due to the interlocutor is. Figure 3.23b shows us how the distribution changes if λ is larger than 2. Here we no longer see a concave distribution for $h = 1$, but already one with several peaks. This is because λ is so large here, that single interactions seriously modify the vocabulary, either towards confirming the old frequencies if h is small and the speaker's own utterances are given more importance, or towards the vocabulary of the interlocutor, if h is large.

Figure 3.24 indicates how the distribution changes for a given h, when the change pressure exerted by an interaction on the vocabulary of a speaker is varied in the parameter range marked by the vertical black lines in Figure

3.4 Utterance Selection Model on a mobile phone network

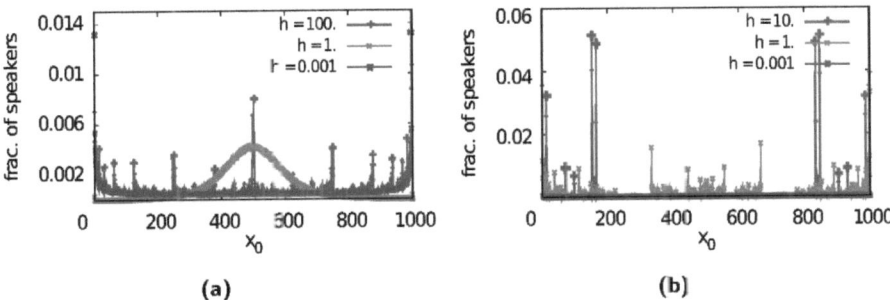

Figure 3.23: Frequency distribution after 1000 loops for $\lambda = 0.01$ and $\lambda = 1$. and different values of h, $r(0) = 0.5$, $T = 1$.

3.22. In (a), for small values of λ, the distribution is centrally peaked, so the vast majority of the speakers use both variants. Increasing λ, the distribution becomes broader, as the changes due to each interaction are more visible. Then, for $h \leq \lambda < 1$, the convex distribution suggests that most of the speakers use only one variant. Finally, for large λ, we obtain a distribution with multiple peaks. Figure 3.24b shows the transition from a concave distribution to one with several peaks, without passing through a convex phase.

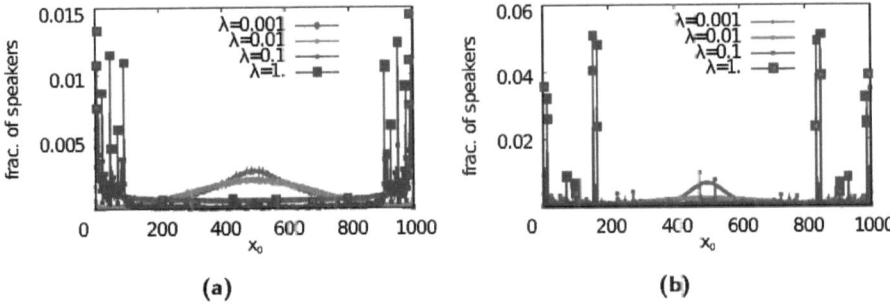

Figure 3.24: frequency distribution after 1000 loops for $h = 0.1$ and $h = 10$. and different values of λ, $r(0) = 0.5$, $T = 1$.

The mobile phone network for which we have explored the quasi-stationary

3. Utterance Selection Model of language change

frequency distribution has, like most social networks [93], a group structure, i.e. agents make phone calls inside some small group for most of the time. To see how the network structure impacts the frequency distribution, we generate a network with randomized topology by randomly rewiring the links. This will dramatically increase the average degree of the nodes and destroy correlations.

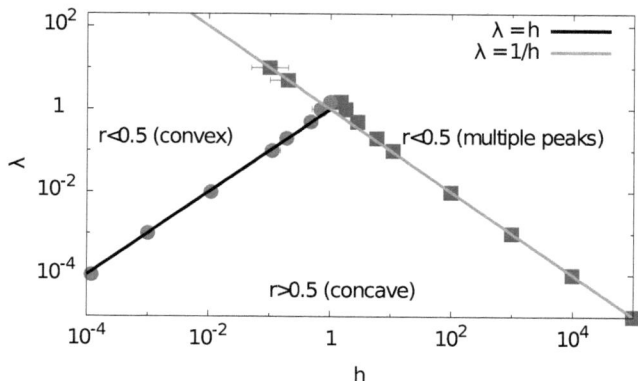

Figure 3.25: (h,λ) diagram of the rewired network. The curves delimiting the regimes of convex, concave and multiple spike distributions are $\lambda = h$ and $\lambda = 1/h$.

Figure 3.25 differs from the (h, λ) diagram of the original network (Figure 3.22) in that the lines marking the boundaries of the three regimes (convex, concave and multiple peak distribution) are shifted: for the network with randomized topology, $\lambda = h$ and $\lambda = 1/h$ are the critical values. Therefore, the clustering of agents in the network has a visible effect on the quasi-stationary frequency distribution.

3.4.2 Outlook

With the aid of algorithms for identifying community structure in a network, like the ones presented by Palla et al. [94], Lancichinetti et al. [95] or Iñiguez

3.4 Utterance Selection Model on a mobile phone network

et al. [96], one could look into the dynamics of the speakers in more detail. For example, it would be interesting to find out whether the fact that very often tightly-knit clusters of speakers interact only weakly with the rest of the network allows these communities to develop their own "dialects", that is, whether in separated parts of the network certain variants are preferred for long periods of time.

Another aspect is the impact of status differences between speakers in such a subdivided population. As we discussed in Section 3.3, if the parameter H_{ij} is heterogeneous for speakers on a complete graph, with increasing width of the distribution consensus is reached faster. However, if the speech community consists of weakly connected groups, individuals with high status inside a group might persuade the rest of the speakers of his group to keep using a variant that is different from the one spoken by the majority of the other speakers, thus impairing global consensus. The influential speakers to whom one can then ascribe higher values of H_{ij} can be located according to the number of calls they make inside their groups.

Since also information about the duration of the phone calls is available in the database, one could use this data to vary the length of the token string uttered by each speaker during an interaction and hereby study the effects of a heterogeneous parameter T on the variant distribution. Also, by shuffling the time stamps of the phone calls, one could verify whether the time order of the interactions plays any role, for instance by altering the quasi-stationary distribution convergence time.

The impact on language change of group structure in the speech community will be discussed in detail with the aid of an extended version of the Utterance Selection Model in the following chapter.

4 Language change in a multiple group society

If speakers coming from distinct backgrounds find themselves united in a group, in time they will develop a common vocabulary in order to communicate successfully [4]. As our society consists of many groups, defined either by geographical location, age, profession or other criteria, we notice two antithetic tendencies that dominate the dynamics of the language: on the one hand, speakers affiliated to a social group will try to reach consensus on a variant in order to describe a particular situation. Since this variant can differ from group to group, another element of rivalry between various forms stems from the interactions between speakers belonging to distinct social groups. Our aim is to understand how the competition among word variants is resolved in such a society composed of several groups of speakers connected with each other. To this end we want to find out how long it takes on average until only one variant is being used throughout the speech community, and which conditions have to be met for consensus to be a realistic outcome.

The Utterance Selection Model gives a good insight into the linguistic dynamics of a group. However, society consists of many groups, with relatively weak connections among them. Hence, to better understand the mechanisms that cause languages to change at word level, we will examine this model in a wider context made up of several interacting non-overlapping groups. Blythe [97] studied the fixation probability and time for a variant in a subdivided

4.1 Multiple Group Utterance Selection Model

population for two different spatial arrangements. For a system in which the groups are well-mixed, the time to consensus is proportional to the number of groups squared. If the groups are placed on a hub-and-spoke network, where in an interaction between groups one of these must be the hub, the fixation time approaches a constant, even in the limit of infinitely many groups. This is because, although the variants of the hub spread much faster the large number of spokes ensures a finite fixation probability of a variant from one of these groups. In our approach, we analyze in detail the influence of increasingly strong separation of the groups on the time that a variant takes to fixate in the whole speech community.

4.1 Multiple Group Utterance Selection Model

In this context, we introduce a new parameter, f, representing the *group affinity*, that is, the probability of a speaker choosing her interlocutor from the same group. $1 - f$ is then the probability that the speaker chooses a conversation partner from another group.

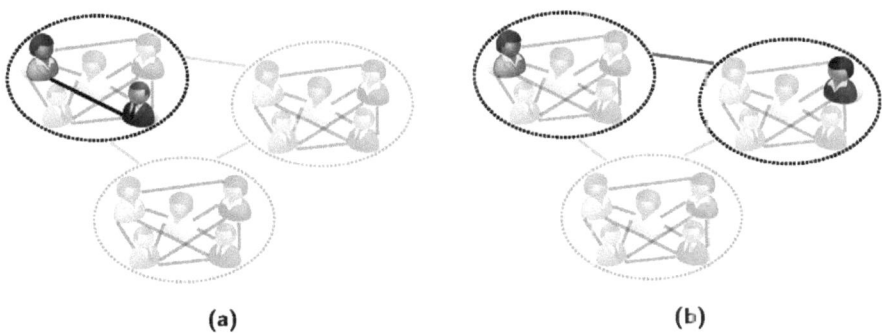

Figure 4.1: (a) The "group affinity" parameter f gives the probability of a speaker to interact with someone from her group. (b) With probability $1 - f$, speaker and interlocutor belong to different groups.

4. Language change in a multiple group society

Restricting the number of variants to two, we can define a measure of consensus in a group x_0 as the average over the first component of the frequency vector **x** for all speakers (the frequency with which the first variant is used in the group), this being a number between zero and one. If x_0 is close to the boundaries of the interval, throughout that group one variant is used for most of the time. If, however, the value of x_0 is close to the center of the interval, speakers use both variants in significant proportions. This does not tell us, however, whether a speaker uses one preferred variant, which differs from speaker to speaker, or all use both variants with comparable frequencies.

Regarding the initial conditions, we will fix half of the groups on one variant and the other half on the other variant. This way, the average time to consensus is larger than for uniformly distributed initial frequencies, because before global consensus on a particular variant can be reached, the variants have to propagate across the groups. Random initial conditions would provide an already shuffled configuration, thus eliminating this mixing time. The time step between interactions, δt, is set to 1 for all simulations presented below.

4.2 Two groups

The first step when moving from one group of speakers to a system composed of many groups is the coupling of two such entities (Figure 4.2).

> "Just imagine two groups living in two neighbouring villages, speaking similar varieties of one language. With the passing of time, their language undergoes constant transformations, but as long as the two communities remain in close contact, their varieties will change in tandem: innovations in one village will soon spread to the other, because of the need to communicate. Now suppose that one of the groups wanders off in search of better land, and loses all contact with the speakers of the other village. The language of the two groups

4.2 Two groups

will then start wandering in different directions, because there will be nothing to maintain the changes in tandem." [3]

Since the further away they are from each other, the less they interact, our parameter f can be seen as the distance between the villages.

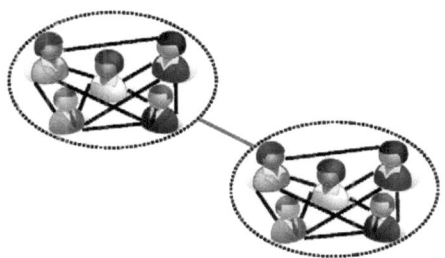

Figure 4.2: Two coupled groups of speakers where speakers interact mostly internally if the group affinity f is large, whereas for low values of f the coupling between the groups becomes increasingly stronger. Each speaker is connected to all others, the links inside the own group having a heavier weight set by the parameter f.

To understand the effect of the coupling, we impose the condition that the two groups are of the same size. Varying the group size N, we numerically investigate the dependence of the average time to consensus on the group affinity parameter f (Figure 4.3) and obtain scaling behavior:

$$t_c(f, N) = N^2 F\Big((1-f)N\Big). \tag{4.1}$$

Here the scaling function F also depends on the number of tokens uttered by each speaker in an interaction, T, the pressure for change on the vocabulary of a speaker, λ, and the relative influence of the interlocutor, h.

There are two asymptotic limits for the time to consensus, which are described by power laws. For weak coupling between groups (large values of f), we find

$$t_c(f, N) \propto N(1-f)^{-1}. \tag{4.2}$$

4. Language change in a multiple group society

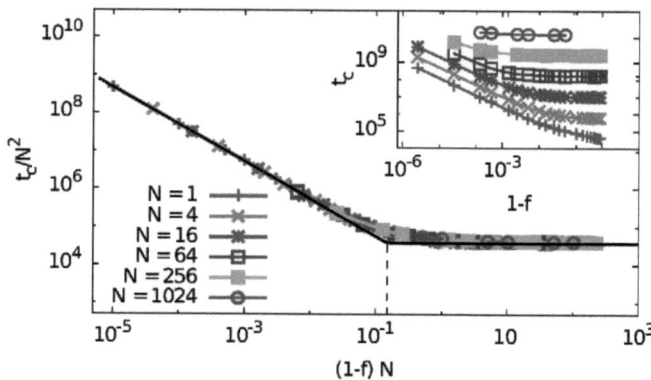

Figure 4.3: Scaling plot of the time to consensus as a function of group size N and group affinity f for two coupled groups. The scaling factors are found as described in Figure 4.4. There are two scaling regimes, for strong and weak coupling respectively. The boundary is marked by the intersection $N_\times(1-f)$ of the curves fitting the power laws. The other parameter values are $T=1, \lambda=0.01, h=0.01$. Inset: The same curves before rescaling.

In contrast, if the coupling is strong, the function F is constant, and therefore the average time to global consensus is given by

$$t_c(f,N) \propto N^2. \tag{4.3}$$

The boundary between these two asymptotic regimes, $N_\times$, defined as the intersection of the above power laws, marks the transition from the one-group to the two-group behavior. While for group sizes $N > N_\times$, one is in the strong coupling regime, the two groups are only weakly coupled for $N < N_\times$. The reason for this is that for the same value of the group affinity f, a smaller group will restore inner consensus faster and thus its language will remain isolated from the one of the other group. In the case of larger groups, the new variant propagates more easily, such that the two groups share both variants for a longer time.

4.2 Two groups

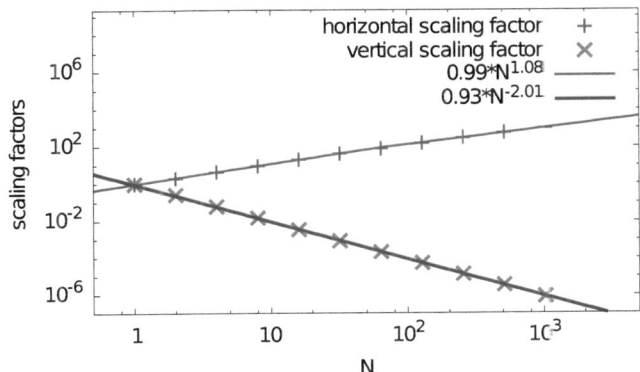

Figure 4.4: To obtain the scaling plot in Figure 4.3 we rescale the curves shown in the inset by shifting them horizontally and vertically by some factor. We plot these factors for various curves in order to find their dependence on the system size N. Disregarding the numerical errors, for the horizontal factor we find a linear dependence on N, so the x-axis of the scaling plot becomes $N(1-f)$, and for the vertical factor a square dependence, hereby turning the y-axis into t_c/N^2.

We find that the value of the crossover group size, $N_\times$, depends on the parameters T, λ and h as follows:

$$(1-f)N_\times = \frac{1}{T}\eta(\lambda, h). \tag{4.4}$$

The result of $N_\times$ being inversely proportional to T, the number of tokens uttered (Figure 4.5), is in accordance with Baxter et al. [1], where they find that T enters the dynamics as a time scale.

Due to the scaling behavior, Eq. (4.1), the location of the crossover point can either be found by varying N for fixed f or vice versa. We choose to keep the group size N fixed and vary the coupling strength f to determine $\eta(\lambda, h)$, which has a complex structure in terms of the parameters λ and h, as can be seen in Figures 4.6 and 4.7.

4. Language change in a multiple group society

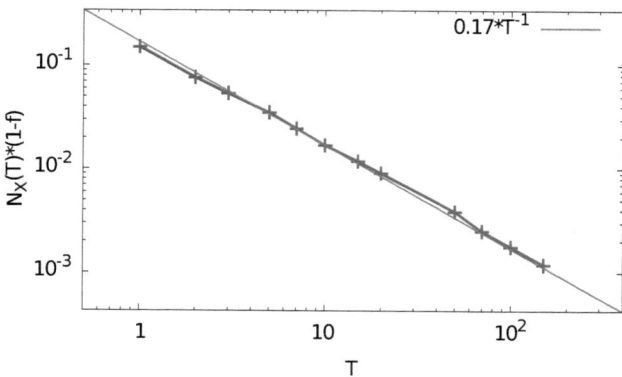

Figure 4.5: The boundary between the asymptotic limits of the scaling law, $(1-f)N_\times$, is inversely proportional to T, the number of uttered tokens. Other parameter values: $N = 2$, $\lambda = 0.01$ and $h = 0.01$.

The parameter λ sets the magnitude of change in vocabulary during an interaction. Investigating the dependence of the function η on λ with the aid of simulations for different values of h, we find that in the limit of small λ, η depends linearly on this parameter. This reflects the fact that for $\lambda \to 0$, when interactions cannot change the vocabularies of the speakers any more, the number of groups makes no difference. On the other hand, for very large λ, η becomes constant, since the update rule (3.3) is now independent of λ, cf. Figure 4.6.

Regarding the role that the interlocutor's influence, h, plays in setting the value of the crossover, we can again characterize the limiting cases. For h very small, one is in a non-interacting regime where the utterances of the interlocutor hardly cause any changes in an agent's vocabulary. The speakers adopt the other variant very reluctantly, and the groups will only adjust their behavior towards each other if they are quite large, so that enough speakers from both groups have the chance to interact. In this limit, η is independent of h, cf. Figure 4.7. In contrast, a large h stands for a very strong influence of the in-

4.2 Two groups

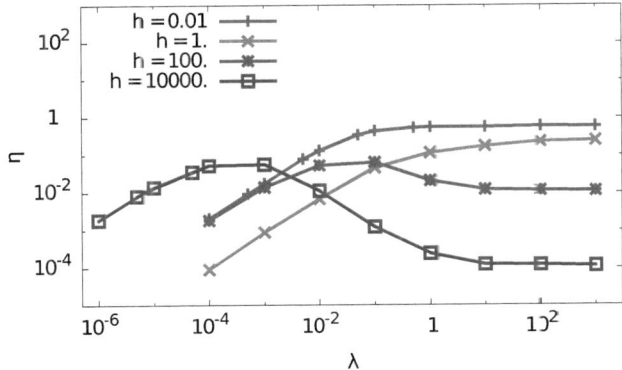

Figure 4.6: The dependence of the boundary between the asymptotic limits of the scaling law, $(1-f)N_\times$, on the magnitude of change in vocabulary λ is linear for $\lambda < 1/h$. For large values of λ, the function η becomes independent of this parameter. Other parameter values are $N = 2$ and $T = 1$.

terlocutor, so that the speaker chooses these utterances as her new vocabulary, and the old frequencies no longer play a role. This leads to speakers easily adopting the variant of an interlocutor from the other group and hereby to a strong coupling between groups. This however does not mean that consensus is reached faster, since the differences in vocabulary between speakers do not decrease significantly. In this last regime, the dependence of η on h is a power law with exponent -1. This means that for $h \to \infty$, $N_\times$ approaches zero. Thus, the dynamics becomes independent of f, and the two groups will behave like one group. We see therefore that a large h can counteract the effect of group segregation. A further observation is that for small λ, η displays a distinct behavior around the point $h = 1$. Approaching this point from below, the interlocutor's influence is large enough to allow the two agents involved in the interaction to align their vocabularies, and thus smaller group sizes are sufficient in ensuring a well-mixed behavior between the groups. At $h = 1$, the utterances of the speaker and her interlocutor have the same weight which enables rapid align-

4. Language change in a multiple group society

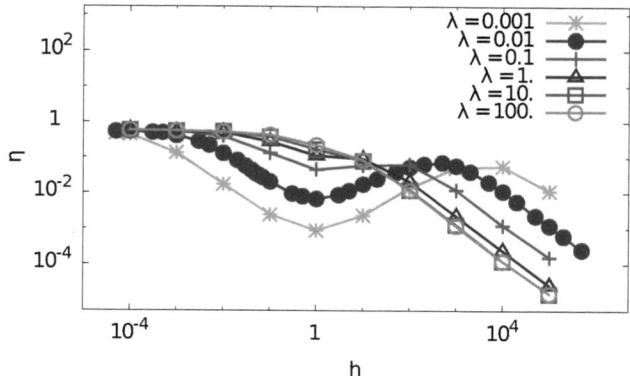

Figure 4.7: For very small values of the the relative influence of the interlocutor, h, the boundary $(1-f)N_\times$ does not depend on this parameter. In the limit of large h, the boundary becomes proportional to $1/h$. The values of the other parameters are $N=2$ and $T=1$.

ment of the speakers' vocabularies through the dissemination of the variants across the groups. For h larger than 1, the relative influence of the speaker's own utterances decreases, and the interlocutor's utterances dominate the interaction. The behavior of η in this parameter range is symmetric with respect to $h=1$.

In the following, we will discuss the weak and strong group coupling regimes in more detail.

4.2.1 Weak coupling

To understand Eq. (4.2), remember that f was the probability for a speaker to interact inside her own group. Then $1-f$ is the probability of an interaction with a speaker from the other group, and

$$\tau := \frac{1}{1-f} \tag{4.5}$$

4.2 Two groups

is the average time between two interactions of this type. If τ is much larger than the average time to consensus in a group, the two groups will evolve independently, each of them reaching internal consensus, and perceive the interactions with the other group only as a series of perturbations (Figure 4.8). Eventually one of the perturbations leads one group to fixate on the variant spoken by the other group. The probability that the group will adopt the variant that the other group has agreed upon is $p = 1/N$ (since in every interaction there are two speakers involved, in a conversation between groups one speaker out of N is "converted" by her interlocutor and then disseminates the opinion in her own group), as can be seen from numerical results in Figure 4.9. The dynamics we are dealing with here is the well-known *gambler's ruin problem* [98].

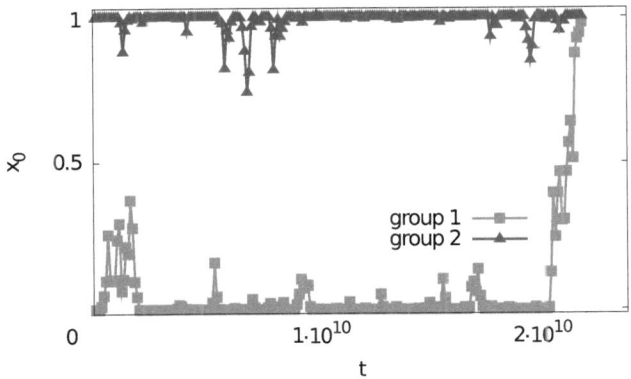

Figure 4.8: The time evolution of the parameter x_0 for a simulation run with $f = 0.9999$ and $N = 256$ (coarse-grained). The two groups notice each other as a series of perturbations. Other parameter values: $T = 1$, $\lambda = 0.01$, $h = 0.01$.

If τ is the average time between inter-group interactions, the number of such interactions until consensus is reached is

$$n \approx \frac{t}{\tau} = t\,(1 - f).$$

4. Language change in a multiple group society

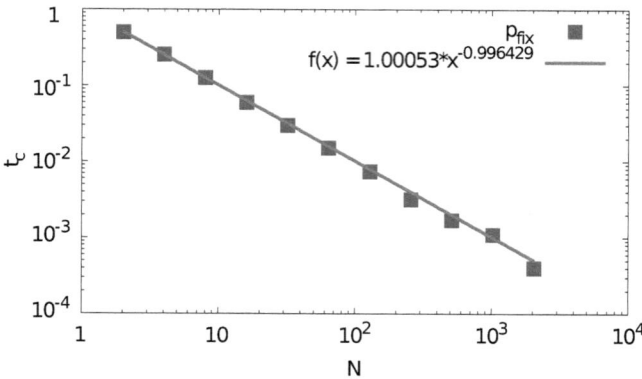

Figure 4.9: The fixation probability of variant 1 (starting with all but one agents speaking variant 0) is equal to $1/N$ in a globally coupled group (*gambler's ruin problem*). Parameter values: $T = 1$, $\lambda = 0.01$, $h = 0.01$, average over 10000 runs.

Out of n trials, the last one is successful, so the probability that the n-th perturbation will lead to global consensus is

$$P(n) = \left(1 - \frac{1}{N}\right)^{n-1} \frac{1}{N}.$$

Since $P(n)$ are the terms of a geometric progression, the probability is properly normalized:

$$\sum_{n=1}^{\infty} P(n) = \frac{1}{1 - (1 - \frac{1}{N})} \cdot \frac{1}{N} = 1. \tag{4.6}$$

4.2 Two groups

The average time to consensus is then given by

$$
\begin{aligned}
t_c = \tau \langle n \rangle &= \tau \sum_{n=1}^{\infty} n P(n) \\
&= \tau \frac{1}{N} \frac{\partial}{\partial x} \sum_{n=1}^{\infty} x^n \Big|_{x=1-\frac{1}{N}} \\
&= \tau \frac{1}{N} N^2 \\
&= \frac{N}{1-f}.
\end{aligned}
\tag{4.7}
$$

4.2.2 Strong coupling

For small values of f, the coupling between the groups is so strong that the groups will evolve towards each other, only to start diffusing together after reaching a common value (Figure 4.10). The two groups thus turn into one large group, a behavior that we have discussed in Chapter 3.

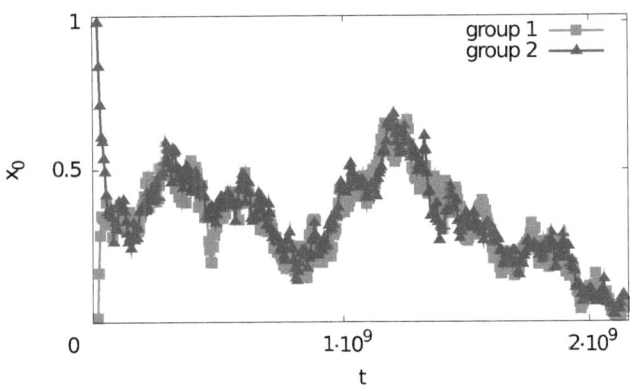

Figure 4.10: A typical time evolution of the parameter x_0 with $f = 0.9$ and $N = 256$ (coarse-grained). The two groups align their vocabularies and evolve together from there on. Other parameter values: $T = 1$, $\lambda = 0.01$, $h = 0.01$.

4. Language change in a multiple group society

Plotting the parameter measuring consensus for each of the two groups as a trajectory in $(x_{0,1}, x_{0,2})$ coordinates (Figure 4.11), the dynamics is that of a biased random walk along the main diagonal of the square, which is quasi one-dimensional. Since there are only two absorbing points, namely $(0,0)$ and $(1,1)$, with increasing system size the "escape windows" become smaller and the time for one of them to be reached (which is the condition for consensus) diverges. This is commonly known as a narrow escape problem [99].

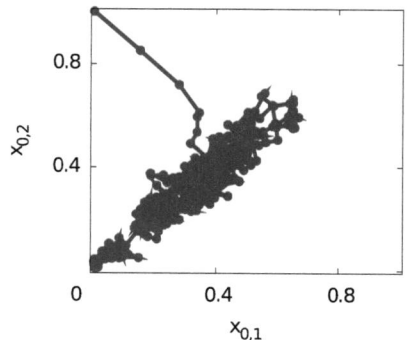

Figure 4.11: The same trajectory as in Figure 4.10 in $(x_{0,1}, x_{0,2})$ coordinates (coarse-grained). The dynamics is that of a one-dimensional random walk. Other parameter values: $T = 1$, $\lambda = 0.01$, $h = 0.01$.

With increasing f, this random walk frays more and more, to the point where it fills out the entire square (Figure 4.13 displays the trajectory shown in Figure 4.12). Here we see that $x_{0,1}(t)$ and $x_{0,2}(t)$ approach each other and separate again. The attraction is a result of the interaction between speakers from the two groups, whereas the seeming repulsion shows that the coupling is not strong enough, since speakers interact more often in their own group.

The two-dimensional narrow escape problem corresponding to Figure 4.13 was treated by Singer et al. [100]. There, the authors solve the problem of a Brownian particle in a rectangular area $\Omega = (0, a) \times (0, b)$ with reflecting boundaries except for a small absorbing interval $\partial \Omega_a = [a - \epsilon, a] \times \{b\}$. The mean first passage time (MFPT) $T(x)$ satisfies the following boundary value

4.2 Two groups

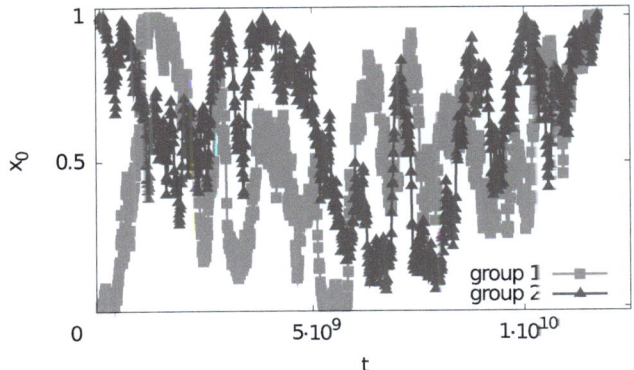

Figure 4.12: A typical time evolution of the order parameter x_0 with $f = 0.995$ and $N = 256$ (coarse-grained) The two trajectories seem to attract and repulse each other every now and then. The latter happens because the coupling is not strong enough to keep the trajectories aligned. Other parameter values: $T = 1$, $\lambda = 0.01$, $h = 0.01$.

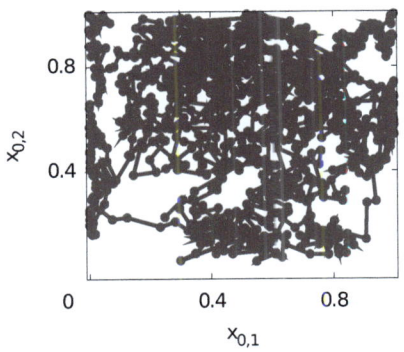

Figure 4.13: The same trajectory as in Figure 4.12 in $(x_{0,1}, x_{0,2})$ coordinates (coarse-grained). The dynamics is that of a two-dimensional random walk in a rectangular area with reflecting boundaries and two absorbing points. The other parameters are: $T = 1$, $\lambda = 0.01$, $h = 0.01$.

problem:

$$\Delta T = -1, \quad (x, y) \in \Omega, \tag{4.8}$$

$$T = 0, \quad (x, y) \in \partial\Omega_a, \tag{4.9}$$

$$\frac{\partial T}{\partial n} = 0, \quad (x, y) \in \partial\Omega - \partial\Omega_a, \tag{4.10}$$

4. Language change in a multiple group society

where Δ is the Laplace operator. The auxiliary function $f = \frac{b^2-y^2}{2}$ is defined, which in turn satisfies

$$\Delta f = -1, \quad (x,y) \in \Omega, \tag{4.11}$$
$$f = 0, \quad (x,y) \in \partial\Omega_a, \tag{4.12}$$
$$\frac{\partial f}{\partial n} = 0, \quad (x,y) \in \{0\} \times [0,b] \cup \{a\} \times [0,b] \cup [0,a] \times \{0\}, \tag{4.13}$$
$$\frac{\partial f}{\partial n} = -b, \quad (x,y) \in [0, a-\epsilon] \times \{b\}. \tag{4.14}$$

Then for the function $u = T - f$ the following equations hold:

$$\Delta u = 0, \quad (x,y) \in \Omega, \tag{4.15}$$
$$u = 0, \quad (x,y) \in \partial\Omega_a, \tag{4.16}$$
$$\frac{\partial u}{\partial n} = 0, \quad (x,y) \in \{0\} \times [0,b] \cup \{a\} \times [0,b] \cup [0,a] \times \{0\}, \tag{4.17}$$
$$\frac{\partial u}{\partial n} = b, \quad (x,y) \in [0, a-\epsilon] \times \{b\}. \tag{4.18}$$

We can write a solution of u as an expansion in eigenfunctions:

$$u(x,y) = \frac{a_0}{2} + \sum_{n=1}^{\infty} a_n \cosh \frac{\pi n y}{a} \cos \frac{\pi n x}{a}, \tag{4.19}$$

with the coefficients a_n to be determined from the boundary conditions at $y = b$:

$$u(x,b) = \frac{a_0}{2} + \sum_{n=1}^{\infty} a_n \cosh \frac{\pi n b}{a} \cos \frac{\pi n x}{a} = 0, \quad x \in (a-\epsilon, a), \tag{4.20}$$

$$\frac{\partial u}{\partial y}(x,b) = \frac{\pi}{a} \sum_{n=1}^{\infty} n a_n \sinh \frac{\pi n b}{a} \cos \frac{\pi n x}{a} = b, \quad x \in (0, a-\epsilon). \tag{4.21}$$

4.2 Two groups

Defining new coefficients $c_n = a_n \sinh\frac{\pi n b}{a}$, these can be rewritten as

$$\frac{c_0}{2} + \sum_{n=1}^{\infty} \frac{c_n}{1 + H_n} \cos n\theta = 0, \quad \pi - \delta < \theta < \pi, \tag{4.22}$$

$$\sum_{n=1}^{\infty} n c_n \cos n\theta = \frac{ab}{\pi}, \quad 0 < \theta < \pi - \delta, \tag{4.23}$$

where $\delta = \frac{\pi \epsilon}{a}$ and $H_n = \tanh(\frac{\pi n b}{a}) - 1$. For $\beta = \exp\{-\frac{\pi b}{a}\} < 1$, $H_n = O(\beta^{2n})$. This problem is mathematically very similar to the one of a Brownian particle confined in an annulus, which can exit only through a narrow region on the inner circle. By solving this problem, Singer et al. find the coefficient c_0 to be

$$c_0 = \frac{2ab}{\pi} \left[2\log\frac{1}{\delta} + 2\log 2 + 4\beta^2 + O(\delta, \beta^4) \right] \tag{4.24}$$

$$= \frac{4ab}{\pi} \left[\log\frac{a}{\epsilon} + \log\frac{2}{\pi} + 2\beta^2 + O(\frac{\epsilon}{a}, \beta^4) \right]. \tag{4.25}$$

The MFPT is then $T(x,y) = u(x,y) + f(y) = \frac{c_0}{2} + \frac{b^2 - y^2}{2}$. Now going back to our problem and inserting $a = N$, $b = N$, taking into account that our diffusion coefficient is $D = 1$ and making use of the initial conditions, the MFPT $T(x,y)$ is given by

$$T(x,y) = \frac{4N^2}{\pi} \left[\log\frac{N}{\epsilon} + \log\frac{2}{\pi} + 2\beta^2 + O(\frac{\epsilon}{a}, \beta^4) \right]. \tag{4.26}$$

Since we are interested in the dependence of the MFPT on N, we can write it as

$$T(x,y) = \frac{4N^2}{\pi} \log\frac{N}{\epsilon} + \text{constant terms}. \tag{4.27}$$

The fact that we are dealing with two absorbing points (symmetric with respect to the main diagonal) instead of one, does not change the N-dependence. Simulations of the simple two-dimensional diffusion process also result in $T(N) \propto N^2 \log N$.

4. Language change in a multiple group society

Being at the crossover of the strong and weak coupling regimes, the range of f for which this dynamics is observed is so narrow that N^2 and $N^2 \log N$ cannot be distinguished in our numerical data. The N^2-dependence however is in good agreement with the simulation results.

4.3 Many groups

Going one step further, we now fix the number of speakers in a group and instead vary the number of interacting groups. Simulations result in a scaling plot, similar to the one for two coupled groups. However, here the underlying phenomena are different. The scaling function for the average time to consensus has the form

$$t_c = N_G^2 \tilde{F}(1-f), \qquad (4.28)$$

where again the function $\tilde{F}$ depends also on the parameters T (the number of tokens uttered by each speaker in an interaction), λ (the magnitude of the vocabulary change due to an interaction), and h (the relative status of the conversation partner). In a system with all-to-all connections between groups and also inside the groups, for strong coupling the one-group result is found again, for the total number of speakers $N \cdot N_G$:

$$t_c \propto (N \cdot N_G)^2. \qquad (4.29)$$

If the groups are weakly coupled, beside the number of groups and the group size, the average time between two interactions that engage speakers from different groups plays an important role:

$$t_c \propto N_G^2 \cdot N \cdot \tau. \qquad (4.30)$$

The square dependence of the time to consensus on the number of groups when

4.3 Many groups

these are placed on a complete graph matches the result of Blythe [97].

On a two-dimensional square lattice, strong coupling leads to the same results as in the well-mixed case, thus t_c is given by Eq. (4.29). For large values of the parameter f, corresponding to weak coupling of the groups, there is a logarithmic correction due to the spatial arrangement of the groups:

$$t_c \propto N_G^2 \cdot \log N_G \cdot N \cdot \tau. \qquad (4.31)$$

The crossover of the two coupling regimes is, as in the case of two groups, found from simulations to be linear in $1/T$ (Figure 4.14), and displays the same complex dependence on the parameters λ and h (Figure 4.15), with a linear dependence on λ for small values of this parameter:

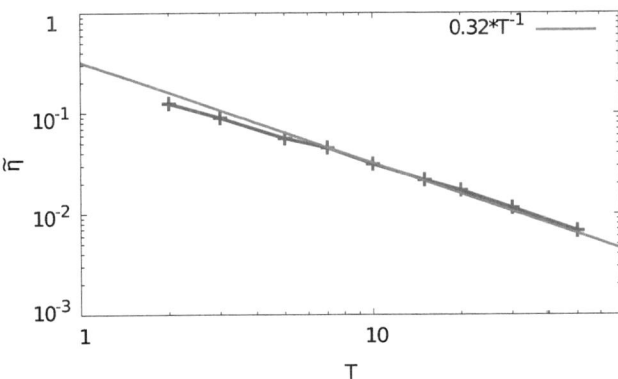

Figure 4.14: The dependence of the crossover $N_{G_\times}$ on T is a power law with exponent -1. $1/T$ is a timescale of the process. Other parameters: N_G, $N = 4$, $\lambda = 0.01$ and $h = 0.01$.

$$N_{G_\times} = \frac{1}{T}\tilde{\eta}(\lambda, h). \qquad (4.32)$$

In the following we will provide a more detailed description of the system's behavior for both the well-mixed and the two-dimensional lattice configurations.

4. Language change in a multiple group society

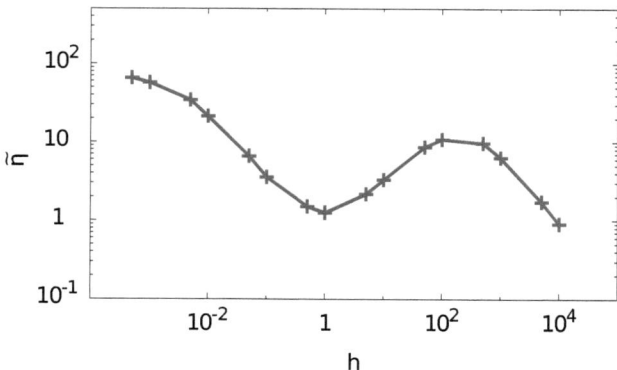

Figure 4.15: The dependence of the boundary $(1-f)N_{G_\times}$ on λ and h displays a complex behavior, discussed for two coupled groups in Section 4.2. Again, $N_G = 4$, $N = 4$, $\lambda = 0.01$ and $T = 1$.

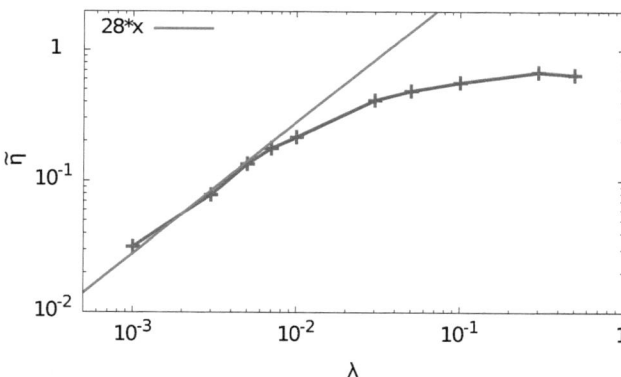

Figure 4.16: The dependence of the crossover group size $N_{G_\times}$ on λ is linear for small values of the parameter. The other parameters are: $N_G = 4$, $N = 4$, $T = 1$ and $h = 0.01$.

4.3.1 Well-mixed system

The simplest instance of a system composed of many connected groups is obtained by placing the groups on the nodes of a complete graph, meaning that

4.3 Many groups

each group interacts with each other group with equal probability. If f is small,

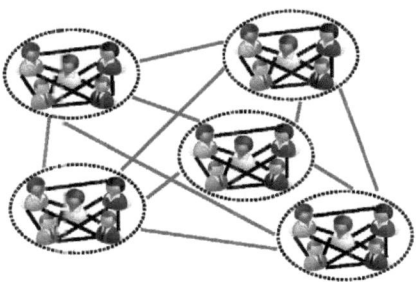

Figure 4.17: Groups of speakers on a well-mixed network. Inside the groups, speakers are also globally connected.

i.e. the coupling between groups is strong, the two variants will diffuse across the groups, both being used in each group for most of the time (Figure 4.18). Again, we recover the time to consensus for one group, i.e.

$$t_c \propto (N \cdot N_G)^2. \tag{4.33}$$

In the regime of large f, which denotes weak coupling, the average time for each of the two groups engaged in an interaction to each achieve inner consensus is very short compared to the time scale τ of interactions between groups. Figure 4.19 shows that the parameters $x_{0,i}$ have either value 0 or 1 for most of the time, indicating the state of inner consensus. The parameter $\langle x_0 \rangle$ however, which measures global consensus, fluctuates as a group changes its inner consensus from one variant to another. When all groups speak the same variant, $\langle x_0 \rangle$ reaches the value 0 or 1 and global consensus is achieved.

When considering interactions between groups, there are three possible outcomes. If before the interaction, both groups shared consensus on the same variant, they remain in this state. If they were using different variants, af-

4. Language change in a multiple group society

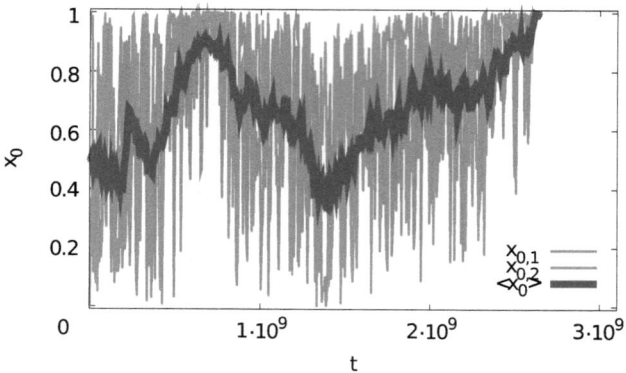

Figure 4.18: A coarse-grained typical trajectory for $f = 0.5$ ($N_G = 256$, $N = 2$, $T = 1$, $\lambda = 0.01$, $h = 0.01$, average over 1000 runs). The parameters $x_{0,1}$ and $x_{0,2}$ represent the average frequency of the first variant in the first and second group (out of the total 256), respectively. $\langle x_0 \rangle$ is the parameter measuring consensus in the entire system.

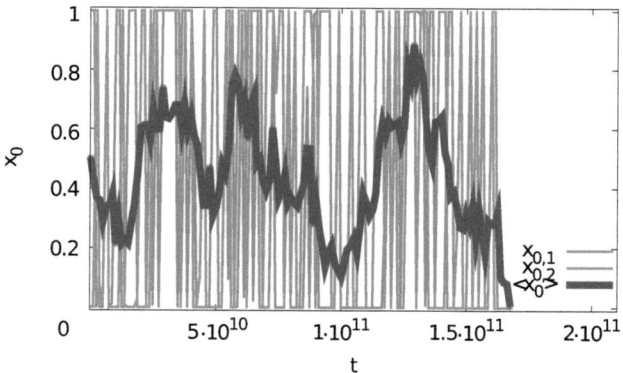

Figure 4.19: A coarse-grained typical trajectory for $f = 0.99$ ($N_G = 256$, $N = 2$, $T = 1$, $\lambda = 0.01$, $h = 0.01$, average over 1000 runs). $x_{0,1}$ and $x_{0,2}$ are the parameters measuring consensus in the first and second group, respectively, whereas $\langle x_0 \rangle$ is the average frequency of the first variant throughout the system.

4.3 Many groups

ter interacting, each of them can switch to the other variant with probability $p = 1/N$ (as discussed already in the two-groups case). If both change to the respective other variant, globally it makes no difference, since the number of groups speaking each variant will be the same as before the interaction. If however only one of the groups changes to the variant of the other, the global balance is shifted towards one of the variants. This behavior corresponds to the Voter Model with link update[1], if time is rescaled so that on average one group opinion change takes place during every inter-group interaction. The stochastic process is a one-dimensional random walk on the interval $[0, N]$ with absorbing boundaries. In the continuous-time description this is given by the diffusion equation

$$\frac{\partial P}{\partial t}(x,t) = D\frac{\partial^2 P}{\partial x^2}. \qquad (4.34)$$

Since consensus is achieved when the random walk reaches 0 or N, the boundaries are absorbing:

$$P(0,t) = P(N,t) = 0. \qquad (4.35)$$

For calculating the mean first passage time of a 1D random walk with absorbing boundaries we follow Gardiner [87]. Let us first solve the general problem: we want to know when a particle that starts at x in the interval $[a, b]$ will leave this interval. The probability for the particle to be in the interval $[a, b]$ is

$$\int_a^b dx' P(x', t|x, 0) \equiv G(x,t). \qquad (4.36)$$

Integrating Eq. (4.34) yields

$$\frac{\partial}{\partial t} G(x,t) = D\frac{\partial^2}{\partial x^2} G(x,t). \qquad (4.37)$$

---

[1] Instead of choosing a node and updating its opinion according to a randomly chosen neighbor, one chooses a link and updates the opinion of one of the two nodes involved. In networks with homogeneous degree distribution (here we have a complete graph, i.e. every speaker is connected to every other one), this choice has no effect on the results.

4. Language change in a multiple group society

The initial condition is a delta function: $P(x,0) = \delta(x)$, which leads to

$$G(x,0) = \begin{cases} \int dx' P(x',0|x,0) = 1, & \text{if } a < x < b \\ 0 & \text{else} \end{cases}. \qquad (4.38)$$

This means that the value of G at the boundaries of the interval $[a, b]$ is

$$G(a,t) = G(b,t) = 0. \qquad (4.39)$$

The mean first passage time is just the time integral of $G(x,t)$:

$$\langle T \rangle = \int_0^\infty dt\, G(x,t). \qquad (4.40)$$

In order to get a differential equation for $T(x)$, we integrate Eq. (4.37) over t. We have

$$\int_0^\infty dt\, \frac{\partial}{\partial t} G = G(x, \infty) - G(x, 0) = -1, \qquad (4.41)$$

and

$$-1 = D \int_0^\infty dt\, \frac{\partial^2}{\partial x^2} G = D \frac{\partial^2}{\partial x^2} T(x), \qquad (4.42)$$

with the boundary conditions

$$T(a) = T(b) = 0. \qquad (4.43)$$

We now want to solve the following equation:

$$\frac{\partial^2}{\partial x^2} T(x) = -\frac{1}{D}. \qquad (4.44)$$

To do that, we first have to find a solution for the homogeneous equation

$$\frac{\partial^2}{\partial x^2} T(x) = 0.$$

4.3 Many groups

This solution is $T_h(x) = \text{const} + \mathcal{N}\int_a^x dy = \text{const} + \mathcal{N}x$. We perform variation of the constant $\mathcal{N} = \mathcal{N}(x)$ to obtain the solution of the inhomogeneous equation. We try the following ansatz, which has been chosen to fulfill the boundary conditions:

$$T(x) = (x-a)\int_x^b dy\, f(y) - (b-x)\int_a^x dy\, f(y), \qquad (4.45)$$

where we have to determine the function f. The derivatives of Eq. (4.45) are

$$\frac{\partial}{\partial x}T = \frac{\partial}{\partial x}T_h(x,a)\int_x^b dy\, f(y) - T_h(x,a)f(x) - \frac{\partial}{\partial x}T_h(b,x)\int_a^x dy\, f(y)$$
$$- T_h(b,x)f(x), \qquad (4.46)$$

$$\frac{\partial^2}{\partial x^2}T = \frac{\partial^2}{\partial x^2}T_h(x,a)\int_x^b dy\, f(y) - 2\frac{\partial}{\partial x}T_h(x,a)f(x) - T_h(x,a)\frac{\partial}{\partial x}f(x)$$
$$- \frac{\partial^2}{\partial x^2}T_h(b,x)\int_a^x dy\, f(y) - 2\frac{\partial}{\partial x}T_h(b,x)f(x) - T_h(b,x)\frac{\partial}{\partial x}f(x). \qquad (4.47)$$

Inserting these into Eq. (4.44), we have

$$-\frac{1}{D} = \underbrace{\frac{\partial^2}{\partial x^2}T_h(x,a)}_{=0}\left[\int_x^b f(y)\,dy + \int_a^x f(y)\,dy\right] - (b-a)\frac{\partial}{\partial x}f(x). \qquad (4.48)$$

Then we are left with

$$\frac{\partial}{\partial x}f(x) = \frac{1}{D(b-a)} \Rightarrow f(x) = \frac{1}{D(b-a)}x. \qquad (4.49)$$

Turning back to Eq. (4.45), we have

85

4. Language change in a multiple group society

$$T(x) = (x-a)\int_x^b \frac{1}{D(b-a)} y\, dy - (b-x)\int_a^x \frac{1}{D(b-a)} y\, dy$$
$$= \frac{1}{2D}(-x^2 + (a+b)x - ab). \qquad (4.50)$$

Now that we obtained the general solution, we substitute $a = 0$, $b = N$ and $D = 1$ and get

$$T(x) = \frac{1}{4}(Nx - x^2). \qquad (4.51)$$

For our initial conditions, $x = \frac{N}{2}$, our simulation results are confirmed: the right part of the curve in Figure 4.3 (left) shows that for small f, $T_c \propto N^2$. On the other hand, if we start close to one of the boundaries (for example $x = 1$), we are dealing with the gambler's ruin problem, and so $T_c \propto N$ (see [101], Chapter 3.2). In general we can thus say, by combining the two previous statements, that the unconditional time to consensus is proportional to the system size N.

On the much shorter time scale of a group changing opinion as a consequence of an interaction with its neighbor, the dynamics inside the group is again the one of the gambler's ruin problem, which we have encountered in the case of two groups with weak coupling. Summing up, the time to reach consensus in the setting of weakly connected groups on a complete graph is

$$t_c \propto N_G^2 \cdot N \cdot \tau. \qquad (4.52)$$

In the scaling plot in Figure 4.20 we see that the left part of the curves for $N_G < 32$ does not overlap with the master curve, but is slightly parallel to it. This is due to finite size effects, since consensus for a small number of groups is reached in a somewhat different way than for large N_G. For $N_G = 2$, global consensus is reached when one of the groups fixates on the variant of the other, so only one change of opinion is needed. If there are more than two groups

4.3 Many groups

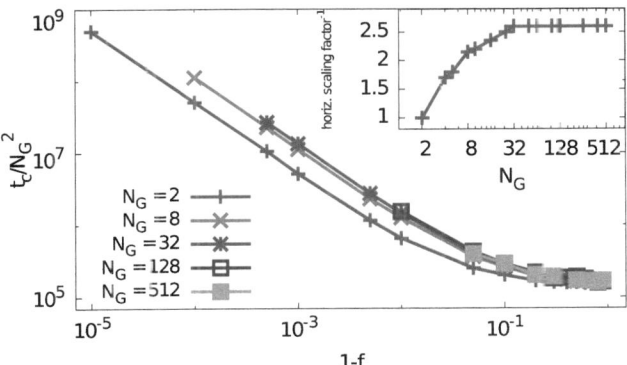

Figure 4.20: Scaling plot of the time to consensus for a well-mixed system with different number of groups and fixed group size. Inset: the horizontal scaling factor obtained by shifting the curves in order to obtain the master curve. For $N_G < 32$, the scaling is different from the one for large N_G due to finite-size effects. Other parameters: $N = 4$, $T = 1$, $\lambda = 0.01$, $h = 0.01$.

involved, a group can change opinion several times before finally all line up, due to interactions with other groups using different variants. As we can read from the inset in Figure 4.20, the many-group behavior is observed for $N_G \geq 32$.

4.3.2 Groups on a lattice

We now place the groups on a square lattice, with each lattice site being occupied by exactly one group. Each group is thus allowed to interact with its four direct neighbors, and the boundary conditions are periodic. As in the well-mixed case, we obtain a scaling plot for the time to consensus (Figure 4.22). For frequent interactions between groups, we have as before

$$t_c \propto (N_G \cdot N)^2. \tag{4.53}$$

4. Language change in a multiple group society

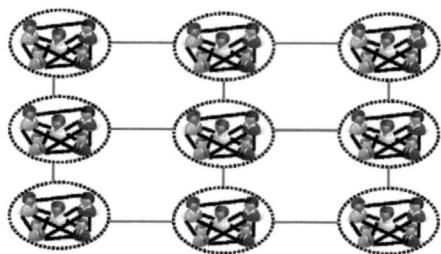

Figure 4.21: Groups of speakers on a square lattice: inside the groups, speakers are well-mixed, each group is connected to its four next neighbors. The boundary conditions are periodic. For an interaction between groups, the speakers are chosen randomly from two neighboring groups.

The difference with respect to the well-mixed case is expressed through a $\log N_G$ correction. This is not surprising, since we are again looking at the Voter Model, this time on a two-dimensional lattice. For this, time to consensus regardless of the initial conditions has been found to be $t_c \propto N_G \log N_G$ ([102, 44]). Imposing the initial conditions that half of the groups start with consensus on one variant and the other half on the other variant, this turns into $t_c \propto N_G^2 \log N_G$ on the time scale τ. Completing the picture with the time needed for a group to change its opinion, proportional to the group size N, the final result is

$$t_c \propto N_G^2 \cdot \log N_G \cdot N \cdot \tau. \qquad (4.54)$$

Thus, as expected, if groups display a spatial arrangement and interact only with their next neighbors, diversity of variants is preserved longer than in a system where all groups can interact with each other.

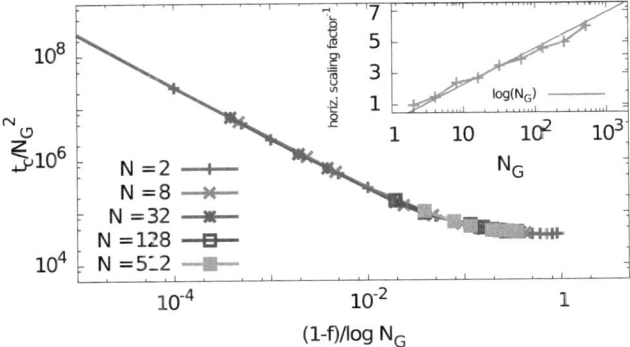

Figure 4.22: Scaling plot of the time to consensus for a system of groups on a square lattice. Inside the groups the configuration is well-mixed. Inset: horizontal scaling factor, obtained by shifting the simulation data curves so that they all fall onto the master curve. Other parameters: $N = 2$, $T = 1$, $\lambda = 0.01$, $h = 0.01$.

4.4 Conclusions

In this chapter we have extended the Utterance Selection Model [1] by giving the underlying social network a more complex structure, allowing for the existence of well-delimited groups inside which speakers interact more often than with the rest of the speech community. We introduced the group affinity f, giving the probability that a speaker chooses his interaction partner to be from the same group, which we used for tuning the strength of the interactions between groups. Our object of interest, the average time until consensus is reached throughout the system, turns out to be highly sensitive to this parameter. Group structure is important, in that it gives rise to various types of behavior, depending on the size and the number of the groups, as well as the status of the interlocutor.

Upon investigating consensus formation in two interacting groups, we obtain a scaling law for the time needed until only one variant is used throughout the speech community. The results tell us that global consensus would be seriously

4. Language change in a multiple group society

impeded if the groups were too large or the interactions between them very scarce. The asymptotic limits of the scaling function show that for strong coupling the entire system behaves like one large group, and global consensus is reached in an average time proportional to the group size N squared. If we further reduce the coupling strength, the average consensus time becomes proportional to the time interval between inter-group interactions, $\tau = 1/(1-f)$, and the group size: $t_c \propto \tau N$. Global consensus is achieved when one of the groups switches to the variant used by the other group, the dynamics corresponding to the gambler's ruin problem.

The boundary between the one-group and the many-group regime has a non-trivial dependence on the parameter h, which represents the influence that the interlocutor's utterances have on the vocabulary with respect to the speaker's own. When this parameter is very small, the speakers ignore each other almost completely, and the variants mix inside a group only if the latter is large enough. With increasing h, speakers start taking into account the utterances of their conversation partner, hereby contributing to the mixing of variants and thus decreasing the critical system size. For very large h, the speakers become "amnesic", meaning that their old vocabulary hardly plays a role any more and they orient their new vocabulary almost entirely after the utterances of the interlocutor. Due to the speakers being highly influenceable, variants spread across the whole system, which behaves like one large group in the limit $h \to \infty$. This means that a large h can counteract even very weak coupling of the groups.

For many coupled groups, a strong connection between them induces a single-group behavior, as all speakers start using both variants, thus again consensus is reached in a time proportional to the total number of speakers squared $(N \cdot N_G)^2$. If groups are more isolated, on the time scale of inter-group interactions the behavior is more complex. Even though groups might reach inner consensus on a variant, they might change their opinion several times, after interacting with

4.4 Conclusions

other groups using different variants, before all agree on one variant and achieve global consensus. This is the dynamics of the voter model with link update for $N_G > 2$, and is quite different from the two-group case, where it was enough for one group to change opinion once. As the number of groups is increased, we observe a finite size effect in the scaling factors. Again, the average time between two interactions between speakers belonging to different groups plays an important role. For a system where the groups are placed on a well-mixed network, the average time to consensus is $t_c \propto N_G^2 \cdot N \cdot \tau$. In the weak coupling limit, the quadratic dependence on the number of groups is owed to the voter model dynamics. Inside a group, one speaker out of N introduces a new variant as a result of an interaction with another group (which takes place on average every τ time steps). Since this dynamics is equivalent to the gambler's ruin problem, a $N \cdot \tau$ term arises. If the groups are positioned on the sites of a square lattice, a logarithmic correction ensues due to the spatial arrangement, and $t_c \propto (N_G^2 \cdot \log N_G) \cdot N \cdot \tau$. In the same way as in the case of two groups, the parameter h controls the position of the boundary between the asymptotic regimes in a complex manner.

We thus learn that not only strong segregation of the various groups, but also excessive partitioning of the speech community can lead to difficulties in reaching global consensus in a realistic period of time. However, 'insecure" speakers, ascribing their interlocutors a much higher importance than themselves and adopting their vocabulary, can accelerate the establishing of a convention.

Latest technological developments are making records of spoken language more and more accessible: the Corpus of Contemporary American English (COCA) [103] contains a large record of digitalized television and radio shows, and offers the tools to compare relative frequencies of words. Following a different approach, New et al. [104] have set up a data base of movie subtitles collected from the Internet and used it to approximate word frequencies in human interactions. Data collections of this type could offer valuable insights for

4. Language change in a multiple group society

future research related to the dynamics of linguistic variants.

5 Viscoelasticity of semiflexible polymer networks with transient cross-links

The cell is the elementary functional unit of all living systems, from unicellular to highly complex organisms. During its lifetime, it reacts to various stimuli and external stress by allowing its shape to undergo reversible modifications. This structural versatility, along with the cell's motility, are accounted for by the cytoskeleton, a network of biopolymer filaments at the same time rigid and flexible, which is also responsible for intracellular transport and cell division. In eukaryotic cells, this polymer framework consists of three types of filaments, namely microtubules, actin filaments and intermediate filaments [105]. In addition to that, a series of proteins are present, which adjust the arrangement of these filaments by cross-linking, capping, bundling them and so on.

Due to their high stability, intermediate filaments provide the cell with mechanical strength; besides that, their role is so far not well understood. Microtubules, in addition to participating in maintaining the cell structure, are of high importance during cell division due to their ability to change conformation, by assembling and disassembling in a very short time. Moreover, they serve as lanes along which molecular motors move within the cell. Finally, the actin filaments are responsible for a large variety of mechanical properties, vital for the efficiency of changes in the shape of the cell as well as for its locomotion. Many of these abilities rely on the viscoelasticity of the biopolymers [106]. Actin binding proteins connect filaments into complex constructs like bundles

5. Viscoelasticity of semiflexible polymer networks

and networks, and enhance the elastic and dissipative properties of the system via the binding dynamics.

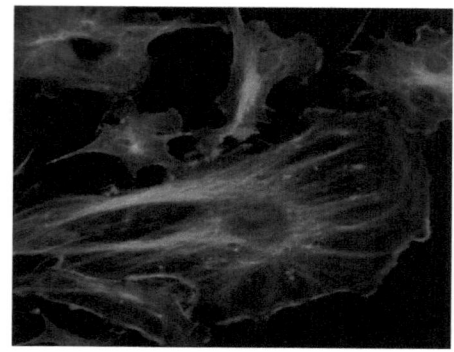

Figure 5.1: Cytoskeleton of the eukaryotic cell: actin filaments are displayed in red and microtubules in green. The nuclei are shown in blue. Image source: Wikipedia.

In the past two decades, single polymer filaments have been the subject of extensive experimental [107, 108, 109] as well as theoretical [110, 111, 112, 113, 114, 115] studies that have succeeded in revealing many of their properties. It was shown [109] that on biologically relevant length scales, actin filaments are described very well by the *worm-like chain model*, which we will introduce in the next section. In networks however, the complex interplay between polymer filaments and the cross-linking proteins is still partly *terra incognita*. In the context where satisfactory descriptions are mostly only available in certain limiting cases, a *bottom-up* approach [116] exploring cell mechanics on various length scales can contribute to shedding more light on the physical properties of semiflexible polymer networks.

As mentioned before, the cytoskeleton is responsible for the mechanical stability of the cell and must therefore be resistant to mechanical forces, but at the same time ensure cell motility and allow for the continuous remodeling of the cellular microstructure. This large variation in functionality can be achieved with the aid of transient cross-links. Actin binding proteins like fascin or heavy mero-myosine (HMM) create this type of non-covalent cross-links in actin networks. These links can open, if the cell is reforming or contracting, and close again, if a stable structure is needed. The proteins can also slide along the filaments, so that a link might not close at the exact same place it was before. This type of cross-links can be characterized by binding and unbinding rates for their opening and closing. The kinetics of transient cross-links can be activated by either force or thermal energy and plays a decisive role in the viscoelastic behavior of the polymer network by triggering a relaxation mechanism, as shown in experiments [117, 118, 119]. Due to the highly nonlinear character of the network, analytical results are very difficult to obtain. Various approaches combining analytic calculations and numerical simulations have aimed at disclosing the shape of the viscoelastic network response to stress [120, 121, 122].

In order to gain some understanding of the role of the binding kinetics of cross-linking proteins in the low-force regime, we will study the linear response of a transiently cross-linked network to a small constant force in the affine deformation regime. This will be accomplished by reformulating the problem in such a way that the complexity of the network is reduced to one polymer filament under the action of a stochastically varying force. Before getting there, we will revisit some known results for the dynamics of single polymer filaments.

5.1 Single polymer dynamics

5.1.1 The worm-like chain model

One of the standard ways of describing a semiflexible polymer theoretically is the *worm-like chain model* (WLC) introduced by Kratky and Porod [123, 124]. The polymer is characterized as an inextensible space curve **r** of length L parametrized by its arc length s, which has a bending stiffness κ. The local curvature of the chain is given by the derivative of the tangent $\mathbf{t}(s) = \partial \mathbf{r}(s)/\partial s$ with respect to s. Bending the polymer requires an energy cost proportional to the square of the curvature.

5. Viscoelasticity of semiflexible polymer networks

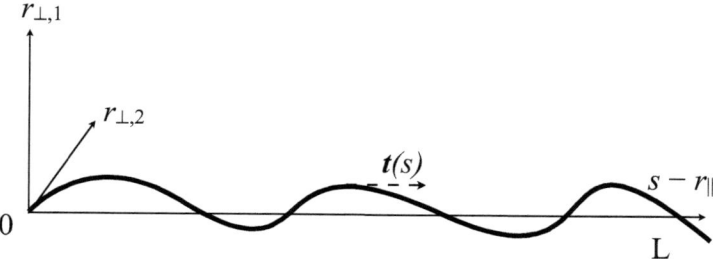

Figure 5.2: The parametrization of a polymer contour with transverse ($r_\perp$) and longitudinal ($r_\parallel$) displacement variables.

The Hamiltonian of the chain can therefore be written as

$$\mathcal{H}_{WLC} = \frac{\kappa}{2} \int_0^L ds \left(\frac{\partial^2 \mathbf{r}}{\partial s^2}\right)^2. \tag{5.1}$$

Together with the local inextensibility condition

$$\left(\frac{\partial \mathbf{r}}{\partial s}\right)^2 = 1, \tag{5.2}$$

Eq. (5.1) provides a complete description of the WLC. Studying this problem analytically is not an easy task, therefore only a few exact results are available. One of them is the tangent correlation function, which decays exponentially:

$$\langle \mathbf{t}(s)\mathbf{t}(s')\rangle = e^{-|s-s'|/l_p}, \tag{5.3}$$

96

5.1 Single polymer dynamics

where $l_p = \kappa/k_B T$ is the *persistence length* of the chain, i.e. a measure for the bending stiffness of the polymer. Depending on the values it takes with respect to the polymer contour length L, it describes stiff ($L \ll l_p$), semiflexible ($L \approx l_p$) and flexible ($L \gg l_p$) polymers. There are however limiting cases which can be studied in more detail, and in the following we will consider a polymer in the weakly bending rod approximation, following Hallatschek et al. [113, 125].

5.1.2 Stochastic equations of motion

Due to the interaction with the solvent molecules in its environment, the conformation of the polymer is subject to fluctuations. Since the worm-like chain is a coarse-grained description of a polymer, on this scale these interactions can be considered random, and hence described stochastically. The Langevin equations of motion are given by the balance of the elastic force $-\delta \mathcal{H}/\delta \mathbf{r}$, the viscous force $\zeta \partial_t \mathbf{r}$ and the thermal noise ξ:

$$\zeta \partial_t \mathbf{r} = -\frac{\delta \mathcal{H}}{\delta \mathbf{r}} + \xi, \tag{5.4}$$

where ζ represents the friction, and the Gaussian white noise correlations are given by the Fluctuation-Dissipation Theorem (FDT) [126]:

$$\langle \xi_n(t) \rangle = 0, \tag{5.5}$$

$$\langle \xi_n(t_1) \xi_m(t_2) \rangle = 4 k_B T \zeta_\perp \delta_{mn} \delta(t_1 - t_2). \tag{5.6}$$

Here, $\zeta_\perp$ represents the friction coefficient for transversal motion of a rod with thickness $a \ll L$ immersed in a solvent with viscosity η:

$$\zeta_\perp = 2\zeta_\parallel \approx \frac{4\pi\eta}{\ln(L/a)}. \tag{5.7}$$

In the weakly bending limit, the transversal and longitudinal deviations from

5. Viscoelasticity of semiflexible polymer networks

a straight line are very small. We express the curve $\mathbf{r}$ with the aid of these:

$$\mathbf{r}(s,t) = (s - r_{\|}(s,t), \mathbf{r}_{\perp}(s,t)), \tag{5.8}$$

where $\mathbf{r}_{\perp}$ is a two-dimensional vector (Figure 5.2). The latter is the reason why in Eq. (5.6) there is a factor 4 instead of 2. Thus we can rewrite the inextensibility constraint Eq. (5.2) as

$$(1 - r'_{\|})^2 + \mathbf{r}'^{\,2}_{\perp} = 1. \tag{5.9}$$

Solving Eq. (5.9) for $r'_{\|}$ and expanding the square root, we obtain the following relation for the derivative of the longitudinal displacement:

$$r'_{\|} = \frac{1}{2}\mathbf{r}'^{\,2}_{\perp} + \mathcal{O}(r'^{\,4}_{\perp}). \tag{5.10}$$

Due to the transverse fluctuations, the end-to-end distance projected onto the longitudinal axis is smaller than the length L of the polymer by an amount $\delta r_{\|} = r_{\|}(L) - r_{\|}(0)$:

$$R_{\|} = L - \delta r_{\|}. \tag{5.11}$$

We can now express $\delta r_{\|}$ in terms of the transverse deviation:

$$\delta r_{\|}(L,t) = \int_0^L \frac{1}{2}\mathbf{r}'_{\perp}(s,t)\,ds. \tag{5.12}$$

The end-to-end distance $R(t)$ in terms of the longitudinal and transverse displacements is

$$R(t) \equiv \left|\int_0^L ds\,\mathbf{r}'\right| \approx R_{\|}(t) + \frac{\Delta \mathbf{r}^2_{\perp}(t)}{2L}, \tag{5.13}$$

with

$$\Delta \mathbf{r}_{\perp} \equiv \mathbf{r}_{\perp}(L,t) - \mathbf{r}_{\perp}(0,t). \tag{5.14}$$

5.1 Single polymer dynamics

Since we can formulate the longitudinal displacement in terms of the transverse one, in the following we need only focus on the latter. The stochastic equations of motion (5.4) then become

$$\zeta \partial_t \mathbf{r}_\perp = -\kappa \mathbf{r}_\perp'''' + \xi_\perp. \tag{5.15}$$

We consider the polymer to have *hinged ends*, thus the appropriate decomposition of $\mathbf{r}_\perp$ is a series of orthonormal sine functions:

$$\mathbf{r}_\perp(s,t) = \sqrt{\frac{2}{L}} \sum_n \mathbf{a}_n(t) \sin(q_n s), \tag{5.16}$$

with modes

$$q_n = \frac{n\pi}{L}. \tag{5.17}$$

In the weakly bending limit, the boundary conditions for hinged ends are:

$$\mathbf{r}_\perp|_{s=0} = \mathbf{r}_\perp|_{s=L} = \mathbf{r}_\perp''|_{s=0} = \mathbf{r}_\perp''|_{s=L} = 0. \tag{5.18}$$

Thus the end-to-end distance $R(t)$ corresponds to the length projected on the longitudinal axis:

$$R(t) \approx R_\parallel(t). \tag{5.19}$$

Inserting the expansion of $\mathbf{r}_\perp$ (5.16) into Eq. (5.15) we obtain the following equation of motion for the amplitude function $\mathbf{a}_n(t)$:

$$\dot{\mathbf{a}}_n = -q_n^4 \mathbf{a}_n + \xi_{\perp,n}. \tag{5.20}$$

This equation is solved by

$$\mathbf{a}_n(t) = \frac{1}{\zeta_\perp} \int_{-\infty}^t \chi_n(t-t') \xi_\perp(t') \, dt', \tag{5.21}$$

5. Viscoelasticity of semiflexible polymer networks

with the Green's function
$$\chi_n(t) = e^{-\kappa q_n^4 t/\zeta_\perp}. \tag{5.22}$$

The longitudinal displacement defined in Eq. (5.12) written in terms of the normal modes becomes

$$\delta r_\|(t) = \frac{1}{L} \int_0^L ds \sum_n \mathbf{a}_n(t) q_n \cos(q_n s) \sum_m \mathbf{a}_m(t) \cos(q_m s). \tag{5.23}$$

Since the mode functions do not depend on s, we only integrate the cosine product, and with a little trigonometry this gives:

$$\int_0^L ds \, \cos(q_n s)\cos(q_m s) = \frac{1}{2}\int_0^L ds \Big(\cos((q_n - q_m)s) + \cos((q_n + q_m)s) \Big)$$

$$= \begin{cases} \dfrac{1}{2}\left(\dfrac{\sin((q_n-q_m)s)}{q_n-q_m}\Big|_0^L + \dfrac{\sin((q_n+q_m)s)}{q_n+q_m}\Big|_0^L \right) = 0 & ,n \neq m, \\ \dfrac{1}{2}\left(\int_0^L dx + \dfrac{\sin((q_n+q_m)s)}{q_n+q_m}\Big|_0^L \right) = \dfrac{L}{2} & ,n = m \end{cases} = \frac{L}{2}\delta_{nm}$$

$$\tag{5.24}$$

Using the above result, we find a simple expression for $\delta r_\|$ in terms of the amplitude functions and the normal modes:

$$\delta r_\|(t) = \frac{1}{L}\sum_{m,n} \mathbf{a}_n(t)\mathbf{a}_m(t) q_n q_m \int_0^L ds \, \cos(q_n s)\cos(q_m s) = \frac{1}{2}\sum_n \mathbf{a}_n^2(t) q_n^2. \tag{5.25}$$

5.1 Single polymer dynamics

5.1.3 Mean square displacement and response function

We are interested in finding the mean square displacement (MSD) of the end-to-end distance:
$$\langle \delta R^2(t) \rangle \equiv \langle \delta[R(t) - R(0)]^2 \rangle. \tag{5.26}$$

In order to achieve this we first calculate the correlation function of the amplitude functions $\mathbf{a}_n$:
$$\langle \mathbf{a}_n(t)\mathbf{a}_m(t') \rangle = \frac{1}{\zeta_\perp^2} \int_{-\infty}^{t} \int_{-\infty}^{t'} \langle \chi_n(t-t_1)\chi_m(t'-t_2) \rangle \langle \xi_n(t_1)\xi_m(t_2) \rangle dt_1 dt_2$$
$$= \frac{2k_B T}{\kappa q_n^4} e^{-\kappa q_n^4 |t-t'|/\zeta}. \tag{5.27}$$

Making use of Wick's theorem [127] to express the four-point correlations in terms of two-point correlations, we get

$$\langle \delta r_\parallel(t)\delta r_\parallel(0) \rangle = \frac{1}{4}\left\langle \sum_n \mathbf{a}_n^2(t)q_n^2 \sum_m \mathbf{a}_m^2(0)q_m^2 \right\rangle = \frac{1}{4}\sum_{n,m} q_n^2 q_m^2 \Big(\langle \mathbf{a}_n^2(t) \rangle \langle \mathbf{a}_m^2(0) \rangle$$
$$+ 2(\langle \mathbf{a}_n(t)\mathbf{a}_m(0) \rangle)^2 \Big)$$
$$= \left(\frac{k_B T}{\kappa}\right)^2 \sum_n q_n^{-4}(1 + 2e^{-2\kappa q_n^4 |t|/\zeta_\perp}). \tag{5.28}$$

The end-to-end correlation function is then found to be
$$\langle \delta r(t)\delta r(0) \rangle - \langle \delta r(0) \rangle^2 = \frac{1}{4}\sum_{n,m} q_n^2 q_m^2 [\langle \mathbf{a}_n^2(t) \rangle \langle \mathbf{a}_m^2(0) \rangle + 2(\langle \mathbf{a}_n(t)\mathbf{a}_m(0) \rangle)^2 - \langle \mathbf{a}_n^2(0) \rangle^2]$$
$$= 2\left(\frac{k_B T}{\kappa}\right)^2 \sum_n q_n^{-4} e^{-2\kappa q_n^4 |t|/\zeta_\perp}. \tag{5.29}$$

5. Viscoelasticity of semiflexible polymer networks

Finally, we find the expression of the MSD:

$$\langle \delta R^2(t) \rangle = \langle (\delta r_\|(0) - \delta r_\|(t))^2 \rangle = 2\langle (\delta r_\|(t))^2 \rangle - 2\langle \delta r_\|(t) \delta r_\|(0) \rangle$$
$$= 4\left(\frac{k_B T}{\kappa}\right)^2 \sum_n q_n^{-4}(1 - e^{-2\kappa q_n^4 |t|/\zeta_\perp}), \qquad (5.30)$$

which at equilibrium becomes

$$\langle \delta R^2 \rangle = 4\left(\frac{k_B T}{\kappa}\right)^2 \sum_n \left(\frac{L}{n\pi}\right)^4 = \frac{2L^4}{45 l_p^2}. \qquad (5.31)$$

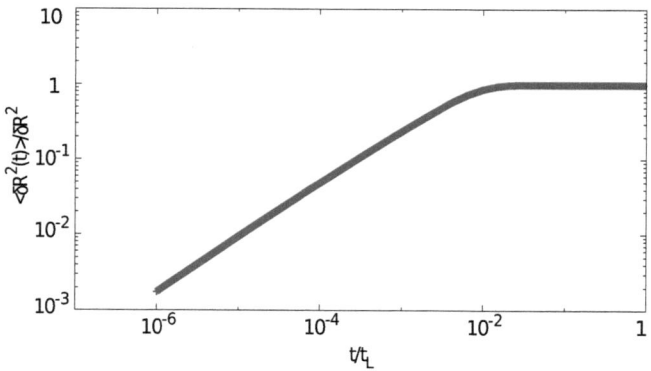

Figure 5.3: Rescaled mean square displacement $\langle \delta R^2(t) \rangle / \langle \delta R^2 \rangle$ as a function of rescaled time $t/t_L^\perp$, with $t_L^\perp = (L/\pi)^4 \zeta/\kappa$ the relaxation time of the longest mode. The scaling of the MSD at times much shorter than $t_L^\perp$ is given by Eq. (5.32).

For times much shorter than the relaxation time of the longest mode, $t_L^\perp = (L/\pi)^4 \zeta/\kappa$, the mean square difference of the end-to-end distance is found [109] to scale as

$$\langle \delta R^2(L, t) \rangle = 0.647 \frac{L}{l_p^2} \left(\frac{t\kappa}{\zeta}\right)^{3/4}. \qquad (5.32)$$

5.1 Single polymer dynamics

Next we calculate the power spectral density by performing a Fourier transform of the end-to-end correlation function given by Eq. (5.29):

$$S(\omega) = 2\left(\frac{k_B T}{\kappa}\right)^2 \int_{-\infty}^{\infty} e^{i\omega t} \sum_n q_n^{-4} e^{-2\kappa q_n^2 |t|/\zeta_\perp} dt$$

$$= 2\left(\frac{k_B T}{\kappa}\right)^2 \sum_n q_n^{-4}\left(\frac{8\kappa/\zeta_\perp}{\omega^2 + 4\kappa^2(\frac{n\pi}{L})^8/\zeta_\perp^2}\right). \quad (5.33)$$

The Fluctuation-Dissipation Theorem gives us the relation between the power spectral density and the imaginary part of the response function [128]:

$$S(\omega) = \frac{2k_B T}{\omega} J''(\omega). \quad (5.34)$$

This way we obtain the following expression:

$$J''(\omega) = \frac{k_B T}{\kappa^2} \sum_n \frac{4\kappa\omega/\zeta_\perp}{\omega^2 + 4\kappa^2(\frac{n\pi}{L})^8/\zeta_\perp}. \quad (5.35)$$

The real part of the response function is found with the aid of the Kramers-Kronig relations [129] (see Appendix (C) for details on the procedure):

$$J'(\omega) = \frac{k_B T}{\zeta^2} \sum_n \frac{8 q_n^{-4}}{\omega^2 \zeta_\perp^2/\kappa^2 q_n^8 + 4}. \quad (5.36)$$

Summing up the results (5.35) and (5.36), the response function has the following form:

$$J(\omega) = J'(\omega) + J''(\omega) = \frac{1}{l_p k_B T}\left(\frac{L}{\pi}\right)^4 \sum_n \frac{1}{n^4 - i\omega\zeta_\perp/(4\kappa(L/\pi)^4)}. \quad (5.37)$$

In [128, 130] the complex viscoelastic modulus of a network on which a macro-

5. Viscoelasticity of semiflexible polymer networks

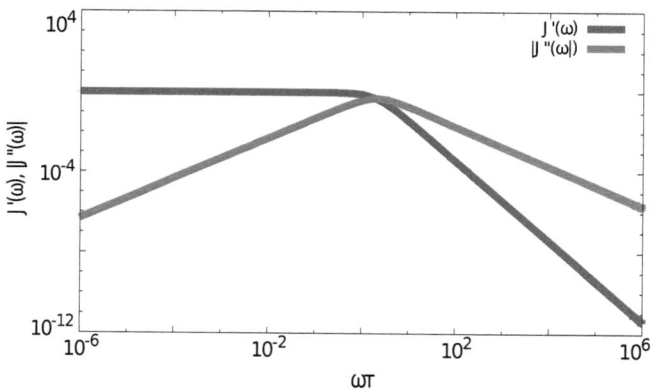

Figure 5.4: Real and imaginary part of the frequency response.

scopic shear strain is applied is given, in terms of the response function of a filament, for a spatial density ρ of filaments:

$$G(\omega) = \frac{1}{15}\rho l_e / J(\omega) - i\omega\eta, \tag{5.38}$$

where $l_e \leq l_p$ is the entanglement length scale and η the viscosity of the solvent. The real part of the shear modulus, the storage modulus $G'(\omega)$, gives the frequency dependent elasticity of the network, whereas the imaginary part, the loss modulus $G''(\omega)$ stands for the viscous dissipation. Taking $L = l_e$, we obtain the plateau of an entangled solution [131]:

$$G^{(0)} = 6\rho k_B T l_p^2 / L^3, \tag{5.39}$$

For high frequencies, we can replace the sum in Eq. (5.37) by an integral, and the response becomes

$$\frac{1}{2\sqrt{2}} \frac{L}{k_B T l_p^2} \left(\frac{2\kappa}{-i\zeta_\perp \omega} \right)^{3/4} \qquad (\omega \ll (\kappa/\zeta)(\pi/L)^4). \tag{5.40}$$

5.2 Affine deformation in a polymer network

Employing Eq. (5.38) again, the shear modulus at high frequencies is proportional to $\omega^{3/4}$ [128, 132]:

$$G(\omega) \approx \frac{1}{15}\rho\kappa l_p(-2i\zeta_\perp/\kappa)^{3/4}\omega^{3/4} - i\omega\eta. \tag{5.41}$$

5.2 Affine deformation in a polymer network

In the affine regime, we make the assumption that the end-to-end distance of each polymer filament follows the macroscopic shear deformation affinely on a coarse-grained scale [133]. Using this argument, one can describe a polymer network with transient cross-links as a set of N_0 parallel filaments with one end fixed and the other able to bind/unbind with rates k_{on}/k_{off}. This situation is similar to the problem of ligand-receptor molecules adhering to a substrate [134, 135, 136].

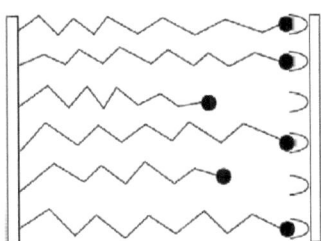

Figure 5.5: Representation of a cross-linked network in the affine deformation regime as N_0 parallel polymer filaments with one end fixed and the other binding/unbinding with rates k_{on}/k_{off} At time t there are $N(t)$ bound filaments.

Given the binding/unbinding rates k_{on} and k_{off}, we want to know the average number of bound polymers at time t. To this end, we set up a master equation

5. Viscoelasticity of semiflexible polymer networks

for the probability that at time t we have N bound filaments:

$$\frac{d}{dt}p_N(t) = k_{on}(N_0-N+1)p_{N-1}(t)+k_{off}(N+1)p_{N+1}(t)-[k_{on}(N_0-N)+k_{off}N]p_N(t). \tag{5.42}$$

The first two terms on the right hand side of Eq. (5.42) represent the gain, i.e. the probability to obtain N bound filaments starting from $N-1$ with an additional one binding, or from $N+1$ with one unbinding. The last two are the loss terms, i.e. the probability that one of the N filaments will bind or unbind, thus changing the number to $N-1$ or $N+1$. Multiplying Eq. (5.42) with N and summing over all possible values we get

$$\sum_{N=1}^{N_0} N \frac{d}{dt}p_N = k_{on}\sum_{N=1}^{N_0}(N_0-N+1)Np_{N-1} + k_{off}\sum_{N=1}^{N_0}(N+1)Np_{N+1}$$
$$-k_{on}\sum_{N=1}^{N_0}(N_0-N)Np_N - k_{off}\sum_{N=1}^{N_0}N^2 p_N, \tag{5.43}$$

where the time dependence has been dropped in order to ease the notation. In the limit of large N_0 shifting the index does not affect the sum, and we obtain

$$\frac{d}{dt}\sum_{N=0}^{N_0} Np_N = k_{on}\sum_{N=0}^{N_0}(N_0-N)p_N - k_{off}\sum_{N=0}^{N_0} Np_N. \tag{5.44}$$

Since $\sum_{N=0}^{N_0} Np_N = \langle N \rangle$ and the probabilities are normalized, we have following differential equation for the average number of bound filaments at time t:

$$\frac{d}{dt}\langle N(t)\rangle = N_0 k_{on} - (k_{on}+k_{off})\langle N(t)\rangle, \tag{5.45}$$

5.2 Affine deformation in a polymer network

which has the solution

$$\langle N(t)\rangle = \langle N(0)\rangle e^{-(k_{on}+k_{off})t} + \frac{N_0 k_{on}}{k_{on}+k_{off}}. \tag{5.46}$$

The equilibrium number of bound polymers is

$$\langle N\rangle = \frac{N_0 k_{on}}{k_{on}+k_{off}}. \tag{5.47}$$

In the same manner we obtain the second moment

$$\langle N^2(t)\rangle = \langle N(0)\rangle^2 e^{-2(k_{on}+k_{off})t} + \frac{(2k_{on}N_0 - k_{on}+k_{off})\langle N(0)\rangle}{k_{on}-k_{off}} e^{-(k_{on}+k_{off})t}$$
$$+ \frac{k_{on}N_0(k_{on}N_0 + k_{off})}{(k_{on}+k_{off})^2}, \tag{5.48}$$

with the equilibrium value

$$\langle N^2\rangle = \frac{k_{on}N_0(k_{on}N_0 + k_{off})}{(k_{on}+k_{off})^2}. \tag{5.49}$$

5.2.1 Mapping of N_0 filaments with constant force onto one polymer with varying force

As discussed in the previous section, in the affine deformation regime we can approximate a network with a set of parallel filaments. This reduces part of the complexity, but in order to describe the dynamics of the network, we still need to solve a system of coupled equations of motion. As this proves to be quite challenging, we have to take into consideration additional ways of simplifying the problem. If we would regard a polymer filament as an ideal spring, then our affine model corresponds to a set of parallel coupled springs. Compared to a system with only one spring, this parallel coupling would only result in a different spring constant, proportional to the number of springs coupled. Introducing

5. Viscoelasticity of semiflexible polymer networks

binding and unbinding of the springs would make the spring constant vary with time. Having a constant force acting on one end of the system, the problem can be reformulated in terms of a varying force acting on a single spring.

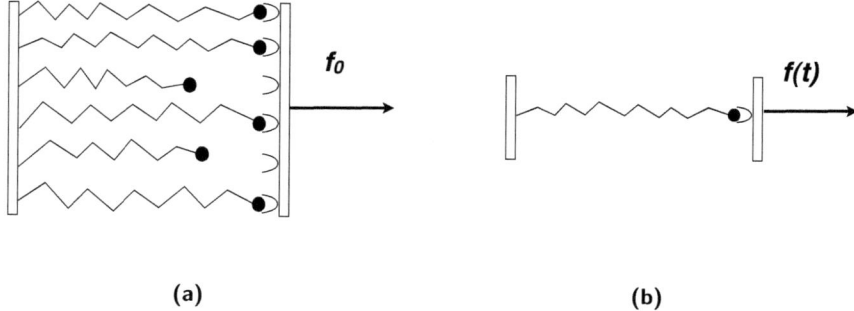

(a) **(b)**

Figure 5.6: (b) A time-varying number of parallel polymers with a constant force acting on them is equivalent to (b) one polymer with a force varying according to the rules of the binding kinetics.

Having thus mapped the bond kinetics onto a random force, we move to a more realistic depiction of a polymer, the worm-like chain, and employ the same idea. Now our task is to analyze the dynamics of a filament subject to Brownian motion and the new force depending on the number of filaments currently bound. Since the force $\mathbf{f_0}$ is small, the number of bound filaments deviates only slightly from the mean number $\langle N(t) \rangle$:

$$N(t) = \langle N(t) \rangle + \delta N(t), \qquad (5.50)$$

where we assume that $\langle \delta N(t) \rangle = 0$. The time correlations for $N(t)$ are

$$\langle N(t)N(t') \rangle = \langle N(t) \rangle^2 + \langle \delta N(t) \delta N(t') \rangle. \qquad (5.51)$$

Making use of the expression for the average number of bound filaments, Eq. (5.47), we get the correlations of $\delta N(t)$:

5.2 Affine deformation in a polymer network

$$\langle \delta N(t) \delta N(t') \rangle = \frac{N_0 k_{on} k_{off}}{(k_{on} + k_{off})^2} e^{-(k_{on}+k_{off})|t-t'|}. \tag{5.52}$$

The force f_N is then expressed in terms of the constant force f_0 and the number $N(t)$ of bound filaments as

$$f_N(t) = \frac{f_0}{N(t)} = \frac{f_0}{\langle N(t) \rangle \left(1 + \frac{\delta N(t)}{\langle N(t) \rangle}\right)} \approx \frac{f_0}{\langle N(t) \rangle} \left(1 - \frac{\delta N(t)}{\langle N(t) \rangle}\right). \tag{5.53}$$

The mean force acting on a polymer filament is

$$\bar{f} = \frac{f_0}{\langle N(t) \rangle} = \frac{f_0 (k_{on} + k_{off})}{N_0 k_{on}}. \tag{5.54}$$

We also need the force correlation function, which is found to be

$$\langle f(t) f(t') \rangle = \left[\frac{f_0(k_{on} + k_{off})}{N_0 k_{on}}\right]^2 + \langle \delta f(t) \delta f(t') \rangle. \tag{5.55}$$

With the assumption that on average the force fluctuations disappear, i.e. $\langle \delta f(t) \rangle = 0$, we have

$$\langle \delta f(t) \delta f(t') \rangle = \frac{\bar{f}^2 k_{off}}{N_0 k_{on}} e^{-(k_{on}+k_{off})|t-t'|} \tag{5.56}$$

We now insert the force f_N given by Eq. (5.53) into the equations of motion for the mode functions, where we thus have two different stochastic processes: first the Gaussian white noise, which was also present in the equations without force (5.21), and the new time-varying force:

$$\zeta_\perp \partial_t \mathbf{a}_n = -(\kappa q_n^4 + \mathbf{f}_N(t) q_n^2) \mathbf{a}_n + \xi. \tag{5.57}$$

The solution of this equation is, as discussed above,

5. Viscoelasticity of semiflexible polymer networks

$$\mathbf{a}_n(t) = \frac{1}{\zeta_\perp} \int_{-\infty}^{t} \chi_n(t-t') \boldsymbol{\xi}_\perp(t') \, dt', \tag{5.58}$$

where the Green's function this time has the following form:

$$\chi_n(t) = e^{-(\kappa q_n^4 t + f_N(t) q_n^2)/\zeta_\perp} \approx e^{(-\kappa q_n^4 + \bar{f} q_n^2)t/\zeta_\perp} \left(1 - \frac{q_n^2}{\zeta_\perp} \delta f\right), \tag{5.59}$$

where we have linearized $\chi_n(t)$ in δf. The time scale for the relaxation of the mode q_n is

$$\tau_n := \frac{\zeta_\perp}{\kappa q_n^4 + \bar{f} q_n^2}. \tag{5.60}$$

Employing the correlations of the Gaussian white noise, Eqs. (5.6), we find the correlations of the mode functions:

$$\langle \mathbf{a}_n(t) \mathbf{a}_m(t) \rangle = \frac{1}{\zeta_\perp^2} \int_{-\infty}^{t} \int_{-\infty}^{t'} dt_1 dt_2 \langle \chi_n(t-t_1) \chi_m(t'-t_2) \rangle \langle \boldsymbol{\xi}_\mathbf{n}(t_1) \boldsymbol{\xi}_\mathbf{m}(t_1) \rangle$$

$$= \frac{4k_B T}{\zeta_\perp} \int_{-\infty}^{t} dt_1 e^{(t+t-2t_1')/\tau_n} \left[1 + \frac{q_n^4}{\zeta_\perp^2} \langle \delta f(t-t_1) \delta f(t'-t_1) \rangle\right]$$

$$= \frac{4k_B T}{\zeta_\perp} e^{-|t-t'|/\tau_n} \left\{\frac{\tau_n}{2} + \frac{q_n^4 \bar{f}^2 k_{\text{off}}}{\zeta_\perp^2 N_0 k_{\text{on}} (k_{\text{on}} + k_{\text{off}})^2} \left[\frac{-\tau_n/2(k_{\text{on}} + k_{\text{off}})}{2/\tau_n + k_{\text{on}} + k_{\text{off}}}\right.\right.$$

$$\left.\left.\cdot \left(e^{-(k_{\text{on}} + k_{\text{off}})|t-t'|} + 1\right) + \frac{\tau_n^2}{2}(k_{\text{on}} + k_{\text{off}})\right]\right\}. \tag{5.61}$$

With this, we can write the MSD of the longitudinal deviation from the contour:

$$\langle \delta R^2(t) \rangle = \left(\frac{k_B T}{\zeta_\perp}\right)^2 4 \sum_n q_n^4 \tau_n^2 \left\{\left[1 + \frac{q_n^4 \bar{f}^2 k_{\text{off}} \tau_n}{\zeta_\perp^2 N_0 k_{\text{on}} (2/\tau_n + k_{\text{on}} + k_{\text{off}})}\right]^2\right.$$

$$- e^{-2t/\tau_n} \left\{1 + \frac{q_n^4 \bar{f}^2 k_{\text{off}}}{\zeta_\perp^2 N_0 k_{\text{on}} (k_{\text{on}} + k_{\text{off}})} \left[-\frac{1}{2/\tau_n + k_{\text{on}} + k_{\text{off}}}\right.\right.$$

$$\left.\left.\left.\cdot \left(e^{-(k_{\text{on}} + k_{\text{off}})t} + 1\right) + \tau_n\right]\right\}^2\right\}. \tag{5.62}$$

5.2 Affine deformation in a polymer network

The end-to-end autocorrelation function is then

$$\langle \delta r(t)\delta r(0)\rangle - \langle \delta r(0)\rangle^2 = 2\left(\frac{k_B T}{\zeta_\perp}\right)^2 e^{-2t/\tau_n}\left\{1 + \frac{q_n^4 \bar{f}^2 k_{\text{off}}}{\zeta_\perp^2 N_0 k_{\text{on}}(k_{\text{on}} + k_{\text{off}})}\right.$$

$$\left.\cdot\left[-\frac{e^{-(k_{\text{on}}+k_{\text{off}})t}+1}{2/\tau_n + k_{\text{on}} + k_{\text{off}}} + \tau_n\right]\right\}^2. \tag{5.63}$$

We can rescale variables so that time becomes dimensionless:

$$\tilde{t} := \frac{t}{\tau}, \qquad \tilde{f} := \frac{\bar{f}}{f_c}, \qquad \tilde{k}_{\text{on}} := k_{\text{on}} \cdot t_L^\perp, \qquad \tilde{k}_{\text{off}} := k_{\text{off}} \cdot t_L^\perp \tag{5.64}$$

where $f_c = \kappa\frac{\pi^2}{L^2}$ is the Euler buckling force, which marks the breakdown boundary of the linear approximation [106], and $t_L^\perp = \zeta/\kappa(L/\pi)^4$ the relaxation time of the longest mode. The MSD becomes, in the rescaled variables:

$$\langle \delta \tilde{R}^2(t)\rangle = \frac{4L^4}{\pi^4 l_p^2}\sum_n \frac{1}{(n^2+\tilde{f})^2}\left\{\left[1 + \frac{\tilde{k}_{\text{off}} n^2 \pi^4 \tilde{f}^2}{2\pi^4 n^2(n^2+\tilde{f}) + \tilde{k}_{\text{on}} + \tilde{k}_{\text{off}}}\right.\right.$$

$$\left.\cdot\frac{1}{N_0 \tilde{k}_{\text{on}}(n^2+\tilde{f})}\right]^2 - e^{-2\pi^4 n^2(n^2+\tilde{f})|\tilde{t}|}\left\{1 + \frac{1}{N_0}\frac{\tilde{k}_{\text{off}}}{\tilde{k}_{\text{on}}}\frac{n^4 \pi^8 \tilde{f}^2}{\tilde{k}_{\text{on}} + \tilde{k}_{\text{off}}}\right.$$

$$\left.\cdot\left[-\frac{1}{2\pi^4 n^2(n^2+\tilde{f}) + \tilde{k}_{\text{on}} + \tilde{k}_{\text{off}}}\cdot\left(e^{-(\tilde{k}_{\text{on}}+\tilde{k}_{\text{off}})|\tilde{t}|}+1\right)\right.\right.$$

$$\left.\left.\left.+\frac{1}{\pi^4 n^2(n^2+\tilde{f})}\right]\right\}^2\right\}. \tag{5.65}$$

Having found this analytical expression, in Figure 5.7 we show this curve together with the MSD for the polymer without force. We see that if the force is smaller than f_c, the two curves cannot be distinguished. Therefore, in the linear response regime, we cannot observe any effect of the binding kinetics.

5. Viscoelasticity of semiflexible polymer networks

Simulations with a bead-rod algorithm [137] performed by Philipp Lang [138] confirm our findings (Figure 5.8).

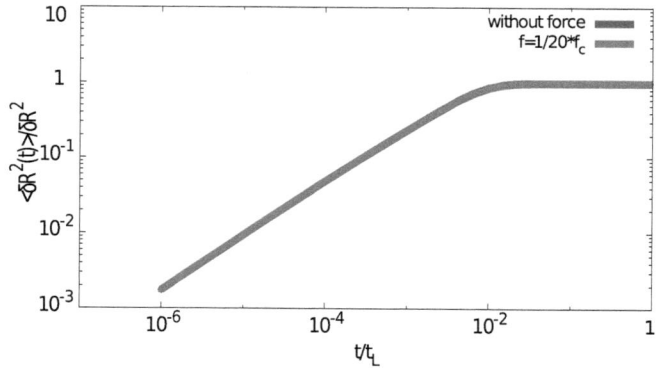

Figure 5.7: Rescaled MSD $\langle \delta R^2(t) \rangle / \langle \delta R^2 \rangle$ as a function of rescaled time for $f = 0$ and $f = 1/50 f_c$, where $f_c = \kappa \pi^2 / L^2$ is the Euler buckling force. There are $N = 10$ cross-links and $k_{\text{on}} = k_{\text{on}} = 0.01$. In the linear response regime there is no visible influence of the binding kinetics on the dynamics.

As for the filament without force (Section 5.1.3), we find the power spectral density by Fourier-transforming the end-to-end correlation function Eq. (5.63):

$$S(\omega) = 2 \left(\frac{k_B T}{\zeta_\perp} \right)^2 \sum_n q_n^4 \tau_n^2 \left\{ \frac{4\tau_n}{\omega^2 \tau_n^2 + 4} \left[1 + \frac{q_n^4 \bar{f}^2 k_{\text{off}}}{\zeta_\perp^2 N_0 k_{\text{on}} (k_{\text{on}} + k_{\text{off}})} \right. \right.$$
$$\left. \cdot \left(\tau_n - \frac{1}{2/\tau_n + k_{\text{on}} + k_{\text{off}}} \right) \right]^2 - \frac{4 \left(2/\tau_n + k_{\text{on}} + k_{\text{off}} \right)}{\omega^2 + \left(2/\tau_n + k_{\text{on}} + k_{\text{off}} \right)^2}$$
$$\cdot \left\{ \frac{q_n^4 \bar{f}^2 k_{\text{off}}}{\zeta_\perp^2 N_0 k_{\text{on}} (k_{\text{on}} + k_{\text{off}})} \cdot \frac{1}{2/\tau_n + k_{\text{on}} + k_{\text{off}}} \right.$$
$$\left. \cdot \left[\frac{q_n^4 \bar{f}^2 k_{\text{off}}}{\zeta_\perp^2 N_0 k_{\text{on}} (k_{\text{on}} + k_{\text{off}})} \left(\frac{1}{2/\tau_n + k_{\text{on}} + k_{\text{off}}} - \tau_n \right) + 1 \right] \right\}$$

5.2 Affine deformation in a polymer network

$$+ \frac{4\left(1/\tau_n + k_{\text{on}} + k_{\text{off}}\right)}{\omega^2 + 4\left(1/\tau_n + k_{\text{on}} + k_{\text{off}}\right)^2} \cdot \frac{q_n^8 \bar{f}^4 k_{\text{off}}^2}{\zeta_\perp^4 N_0^2 k_{\text{on}}^2 (k_{\text{on}} + k_{\text{off}})^2}$$

$$\cdot \frac{1}{\left(2/\tau_n + k_{\text{on}} + k_{\text{off}}\right)^2} \Bigg\}. \tag{5.66}$$

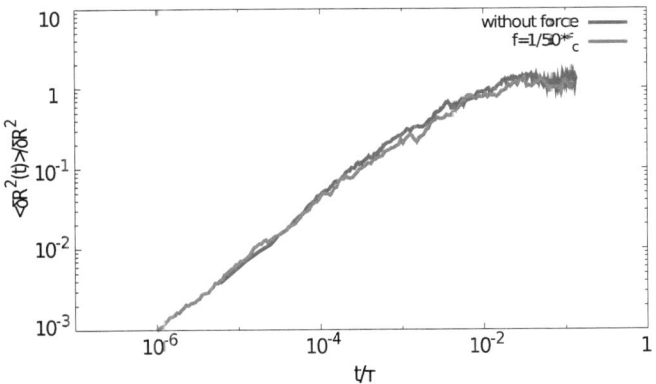

Figure 5.8: Rescaled MSD $\langle \delta R^2(t) \rangle / \langle \delta R^2 \rangle$ as a function of rescaled time for the cases with and without force, obtained from simulations with a bead-rod algorithm for $L = l_p = 20$ and the force $f = 1/50 f_c$. There are N=10 cross-links, $k_{\text{on}} = k_{\text{on}} = 0.01$. As for the analytic results (Figure 5.7), there is no visible difference between the two curves.

The imaginary part of the response function is

$$J''(\omega) = \frac{k_B T}{\zeta_\perp^2} \sum_n q_n^4 \tau_n^2 \Bigg\{ \frac{4\omega \tau_n}{\omega^2 \tau_n^2 + 4} \Bigg[1 + \frac{q_n^4 \bar{f}^2 k_{\text{off}}}{\zeta_\perp^2 N_0 k_{\text{on}} (k_{\text{on}} + k_{\text{off}})} \\ \cdot \left(\tau_n - \frac{1}{2/\tau_n + k_{\text{on}} + k_{\text{off}}} \right) \Bigg]^2 - \frac{4\omega \left(2/\tau_n + k_{\text{on}} + k_{\text{off}}\right)}{\omega^2 + \left(2/\tau_n + k_{\text{on}} + k_{\text{off}}\right)^2}$$

5. Viscoelasticity of semiflexible polymer networks

$$\cdot \left\{ \frac{q_n^4 \bar{f}^2 k_{\text{off}}}{\zeta_\perp^2 N_0 k_{\text{on}}(k_{\text{on}} + k_{\text{off}})} \cdot \frac{1}{2/\tau_n + k_{\text{on}} + k_{\text{off}}} \right.$$

$$\cdot \left[\frac{q_n^4 \bar{f}^2 k_{\text{off}}}{\zeta_\perp^2 N_0 k_{\text{on}}(k_{\text{on}} + k_{\text{off}})} \left(\frac{1}{2/\tau_n + k_{\text{on}} + k_{\text{off}}} - \tau_n \right) + 1 \right]$$

$$+ \frac{4\omega \left(1/\tau_n + k_{\text{on}} + k_{\text{off}}\right)}{\omega^2 + 4\left(1/\tau_n + k_{\text{on}} + k_{\text{off}}\right)^2} \cdot \frac{q_n^8 \bar{f}^4 k_{\text{off}}^2}{\zeta_\perp^4 N_0^2 k_{\text{on}}^2 (k_{\text{on}} + k_{\text{off}})^2}$$

$$\left. \cdot \frac{1}{\left(2/\tau_n + k_{\text{on}} + k_{\text{off}}\right)^2} \right\}. \tag{5.67}$$

Employing the Kramers-Kronig relations [129] we also find the real part of the response function (calculations presented in Appendix C).

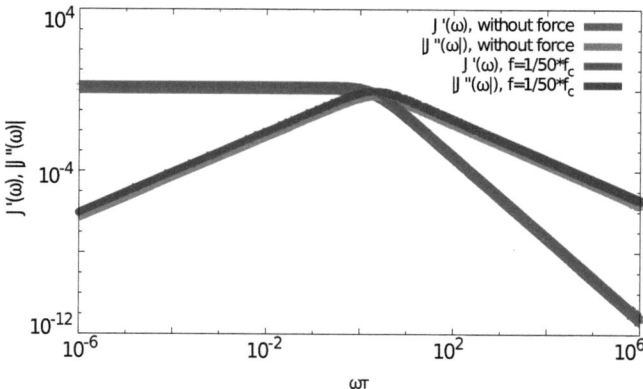

Figure 5.9: Real ($J'(\omega)$) and imaginary part ($J''(\omega)$) of the frequency response for the cases with and without force.

$$J'(\omega) = \frac{k_B T}{\zeta_\perp^2} \sum_n q_n^4 \tau_n^2 \left\{ \frac{8}{\omega^2 \tau_n^2 + 4} \left[1 + \frac{q_n^4 \bar{f}^2 k_{\text{off}}}{\zeta_\perp^2 N_0 k_{\text{on}}(k_{\text{on}} + k_{\text{off}})} \right] \right.$$

$$\cdot \left(\tau_n - \frac{1}{2/\tau_n + k_{\text{on}} + k_{\text{off}}} \right) \right]^2 - \frac{4 \left(2/\tau_n + k_{\text{on}} + k_{\text{off}} \right)}{\omega^2 + \left(2/\tau_n + k_{\text{on}} + k_{\text{off}} \right)^2}$$

$$\cdot \left\{ \frac{q_n^4 \bar{f} k_{\text{off}}}{\zeta_\perp^2 N_0 k_{\text{on}}(k_{\text{on}} + k_{\text{off}})} \left[1 + \frac{q_n^4 \bar{f}^2 k_{\text{off}} \tau_n \left(1/\tau_n + k_{\text{on}} + k_{\text{off}}\right)}{\zeta_\perp^2 N_0 k_{\text{on}}(k_{\text{on}} + k_{\text{off}}) \left(2/\tau_n + k_{\text{on}} + k_{\text{off}} \right)} \right] \right.$$

$$+ \frac{4 \left(1/\tau_n + k_{\text{on}} + k_{\text{off}}\right)^2}{\omega^2 + 4\left(\frac{1}{\tau_n} + k_{\text{on}} + k_{\text{off}}\right)^2} \cdot \frac{q_n^8 \bar{f}^4 k_{\text{off}}^2}{\zeta_\perp^4 N_0^2 k_{\text{on}}^2 (k_{\text{on}} + k_{\text{off}})^2}$$

$$\left. \cdot \frac{1}{\left(2/\tau_n + k_{\text{on}} + k_{\text{off}} \right)^2} \right\}. \tag{5.68}$$

5.3 Conclusions

Studying a transiently cross-linked semiflexible polymer network in the affine deformation regime allows us to make simplifying assumptions so that we can view the network as a set of parallel filaments fixed at one end, with the other end unbinding and binding again as the cross-links open and close due to thermal noise or external forces. Employing the *worm-like chain* description of a semiflexible polymer filament, we can formulate a set of coupled equations of motion for this system, which is however too complex to enable the analytic calculation of the linear response function.

In this chapter we have introduced a considerable simplification to the affine network model. Assuming that the force acting on the the system is distributed evenly on all filaments, the binding and unbinding events of the latter result in a varying force on each bound filament. Thus we mapped the problem

5. Viscoelasticity of semiflexible polymer networks

of many parallel polymers binding and unbinding while subject to a constant external force onto a single polymer with a force that varies according to the binding kinetics. This way, we only have to solve the equation of motion for one filament in order to find the linear response function of the network with binding/unbinding events. However, in order to remain in the linear regime and thus ensure the validity of the above mentioned assumptions, we have to consider forces that are much lower than the Euler buckling force $f_c = \kappa \pi^2 / L^2$.

The results of our calculations suggest that the response function in the case with force is hardly different from the one without force, as can be seen from the plots presented above (Figures 5.7 and 5.9), where the curves overlap completely. Simulations using a bead-rod algorithm, where a stochastically changing force acts upon a polymer filament modeled as a series of beads connected by stiff rods, result in a similar picture (Figure 5.8), which agrees very well with our findings. This means that the forces taken into consideration are too small to cause a significant contribution of the binding behavior to the stress relaxation of the network.

Approaches like the ones of Heussinger [121] or Wolff et al. [122] study the nonlinear response of a transiently cross-linked network with a combination of analytic calculations and numerical simulations. In [121], the author describes the cross-links as nonlinear elastic elements which significantly increase the stiffness of the network when subject to stress and considers the interplay between the elastic properties of the filaments and the cross-links. It results that the stiffness of the cross-links plays an important role in stress relaxation when it is of the same magnitude as the filament stiffness and the unbinding of cross-links is frequent. Wolff et al. revert to the *glassy worm-like chain* [139], which models the interactions between filaments by an exponential stretching of the single filament relaxation times. An incorporated prestressing force accounts for the network deformation and also influences the binding rates of the cross-links. In the context of the inelastic glassy worm-like chain, the stiffening of the network

5.3 Conclusions

is owed to the nonlinear stretch response of the filaments, whereas the softening after an applied strain is the result of the slow bond kinetics. The interplay of these two components leads to a complex viscoelastic behavior on multiple time scales.

We conclude that although the affine model has the merit of allowing the analytical treatment of the viscoelastic response for a network with cross-linker binding and unbinding events, it is not sufficient for reproducing the viscoelastic behavior observed in various experiments [117, 118, 119]. Therefore, in order to obtain a visible effect of the external force on the binding kinetics, one has to allow for non-affine deformations of the network and investigate the nonlinear response in a wider force range.

A Generating uncorrelated scale free random networks

Social networks are scale-free and display degree-degree correlations, since nodes with high connectivity are usually linked to other well-connected nodes [140]. However, when studying such a network, in most cases analytical results can only be obtained if correlations are absent. This is why uncorrelated networks are important as null models for testing various properties of dynamic systems. Scale-free networks are described by a power-law degree distribution:

$$P(k) \sim ck^{-\gamma}, \tag{A.1}$$

where γ is a characteristic degree exponent. For the most frequent scale-free networks in nature, $2 < \gamma \leq 3$.

An algorithm for generating uncorrelated scale-free random networks is presented by Catanzaro et al. [141]. The authors find that in order to have truly uncorrelated networks, the maximum degree of a node has to be $N^{1/2}$. In this algorithm, each vertex i out of the N composing the network is assigned a number k_i of "stubs", i.e. parts of edges that will subsequently form pairs to connect two nodes with each other. These stubs are drawn from the distribution $P(k)$ with $m \leq k \leq N^{1/2}$, where m is the minimal degree of a node, and as mentioned above, $N^{1/2}$ is the maximum degree. Also, the condition is imposed that $\sum_i k_i$ be even. The stubs are then connected two by two, making sure that they do not belong to the same node, or that two nodes are linked by more than one edge. In order to respect these constraints, Catanzaro et al. suggest an implementation where, after the stubs have been assigned to each node, a list of $\sum_i k_i$ elements is created, with k_i instances of node i. Next, two elements are chosen. If they belong to the same node or a link between those two nodes already exists, other two elements are drawn. Otherwise, the link is formed and the pair is removed from the list. This is repeated until the list is empty.

To make sure that the network generated in this way consists of one connected component, we test it with the algorithm of Tarjan [142]. A depth-first search is started from an arbitrary node, each node v being placed on a stack and assigned a running number v.index in the order they are discovered. In addition to that, each node is assigned a value v.lowlink which is equal to the smallest index of a node reachable from v. This value is always smaller than the index of v, or equal to it, if v is not connected to any vertex with lower index. The value v.lowlink helps finding the root of the connected component (i.e. the first node encountered during the search that belongs to the component), since for the root, v.index=v.lowlink. When the search in a subtree is finished, the nodes are removed from the stack and each node is checked on whether it is the root of a connected component. If this is the case, this node and all the nodes removed from the stack before it form the strongly connected component. If there are nodes of the network that do not belong to this component, a new depth-first search is started in a node that has not yet been visited. The procedure is repeated until every node in the network is found in a connected component. Since for our purposes we need networks consisting of one connected component, if the search results in several of these, they are connected among each other through a random link between nodes with degree less than $N^{1/2}$. This procedure should not introduce correlations, since the conditions above are respected.

B Gillespie algorithm

To simulate diffusion on a two-dimensional square lattice in Chapter 3.3.3, we employ a simplified version of the Monte Carlo algorithm developed by Gillespie [143, 144], an exact method for simulating N chemical species engaged in M types of coupled reactions. If the simulation time step is much smaller than the time between two reactions, it is more efficient to estimate the time when the next reaction will occur and update time accordingly, rather than checking at every time step whether there is any due reaction. For this, the reaction probability density function $P(\tau, \mu)$ is introduced, where $P(\tau, \mu)d\tau$ is the probability that, given the state of the system at time t, the next reaction will take place in the time interval $(t+\tau, t+\tau+d\tau)$ and will be of type μ. If c_μ (with $\mu \in \{1..M\}$) are the reaction rates for the M types of chemical reactions, then the average probability that a reaction of type μ will take place in the next time interval $(t, t+dt)$ is $c_\mu \, dt$. Then, if h_μ is the number of combinations of distinct molecular reactants that can participate in a reaction of type μ, the probability of a reaction of this type occurring in the time interval $(t, t+dt)$ given the state of the system at time t is $h_\mu c_\mu \, dt$. The above defined probability for a reaction of type μ to occur in the interval $(t+\tau, t+\tau+dt)$ is given by

$$P(\tau, \mu)d\tau = P_0(\tau) h_\mu c_\mu \, d\tau, \tag{B.1}$$

where $P_0(\tau)$ is the probability that no reaction will take place in the interval $(t, t+\tau)$, given the state of the system at time t. $1 - \sum_\nu h_\nu c_\nu d\tau'$ is the probability that no reaction will occur in the interval τ' from this particular state of the system. Then we have

$$P_0(\tau' + d\tau') = P_0(\tau')\left(1 - \sum_{\nu=1}^{M} h_\nu c_\nu d\tau'\right), \tag{B.2}$$

from which

$$P_0(\tau) = \exp\left[-\sum_{\nu=1}^{M} h_\nu c_\nu \tau\right]. \tag{B.3}$$

Inserting this expression into (B.1) results in

$$P(\tau, \mu) = \begin{cases} h_\mu c_\mu \exp\left[-\sum_{\nu=1}^{M} h_\nu c_\nu \tau\right] & \text{if } 0 \leq \tau < \infty \\ 0 & \text{else} \end{cases}. \tag{B.4}$$

Finally, to find the time τ at which the reaction of type μ will take place, we generate two random numbers, r_1 and r_2, from the unit interval uniform distribution, with which we obtain

$$\tau = 1 \Big/ \left(\sum_{\nu=1}^{M} h_\nu c_\nu\right) \ln 1/r_1, \tag{B.5}$$

and μ as the integer for which the following relation holds:

$$\sum_{\nu=1}^{\mu-1} h_\nu c_\nu < r_2 \sum_{\nu=1}^{M} h_\nu c_\nu \leq \sum_{\nu=1}^{\mu} h_\nu c_\nu. \tag{B.6}$$

Since in our model we only have one "chemical species", the speakers, and one type of reaction, which is the swapping of two individuals with rate d, things become simpler, since we only have to calculate when the next diffusion step will take place. We generate a random number r and find

B. Gillespie algorithm

$$\tau = 1/d \; \log 1/r \; dt, \tag{B.7}$$

update the current time to $t + \tau$ and start a new step of the algorithm.

As a final remark, the Gillespie algorithm is similar in spirit to the BKS method for simulating Ising spin systems at low temperatures [145].

C Calculating the real part of the response function

For a complex function $J(\omega) = J'(\omega) + iJ''(\omega)$, analytic in the upper half plane of ω and which vanishes faster than $\frac{1}{\omega}$ as $|\omega| \to \infty$, the Kramers-Kronig relations are given by:

$$J'(\omega) = \frac{1}{\pi} \mathcal{P} \int_{-\infty}^{\infty} \frac{J''(\omega')}{\omega' - \omega} d\omega' \tag{C.1}$$

and

$$J''(\omega) = -\frac{1}{\pi} \mathcal{P} \int_{-\infty}^{\infty} \frac{J'(\omega')}{\omega' - \omega} d\omega' , \tag{C.2}$$

where $\mathcal{P}$ represents the Cauchy principal value. The imaginary part of the response function, found in Chapter 5, is

$$J''(\omega) = \frac{k_B T}{\zeta_\perp^2} \sum_n q_r^2 \tau_n^2 \Bigg\{ \frac{4\omega \tau_n}{\omega^2 \tau_n^2 + 4} \left[1 + \frac{q_n^4 \bar{f}^2 k_{\text{off}}}{\zeta_\perp^2 N_0 k_{\text{on}}(k_{\text{on}} + k_{\text{off}})} \right.$$

$$\left. \cdot \left(\tau_n - \frac{1}{2/\tau_n + k_{\text{on}} + k_{\text{off}}} \right) \right]^2 - \frac{4\omega \left(2/\tau_n + k_{\text{on}} + k_{\text{off}}\right)}{\omega^2 + \left(2/\tau_n + k_{\text{on}} + k_{\text{off}}\right)^2}$$

$$\cdot \left\{ \frac{q_r^2 \bar{f}^2 k_{\text{off}}}{\zeta_\perp^2 N_0 k_{\text{on}}(k_{\text{on}} + k_{\text{off}})} \cdot \frac{1}{2/\tau_n + k_{\text{on}} + k_{\text{off}}} \right.$$

$$\left. \cdot \left[\frac{q_n^4 \bar{f}^2 k_{\text{off}}}{\zeta_\perp^2 N_0 k_{\text{on}}(k_{\text{on}} + k_{\text{off}})} \left(\frac{1}{2/\tau_n + k_{\text{on}} + k_{\text{off}}} - \tau_n \right) + 1 \right] \right\} \Bigg\}$$

C. Calculating the real part of the response function

$$
\begin{aligned}
&+ \frac{4\omega\left(1/\tau_n + k_{\text{on}} + k_{\text{off}}\right)}{\omega^2 + 4\left(1/\tau_n + k_{\text{on}} + k_{\text{off}}\right)^2} \cdot \frac{q_n^8 \bar{f}^4 k_{\text{off}}^2}{\zeta_\perp^4 N_0^2 k_{\text{on}}^2 (k_{\text{on}} + k_{\text{off}})^2} \\
&\quad \cdot \frac{1}{\left(2/\tau_n + k_{\text{on}} + k_{\text{off}}\right)^2} \\
&= \frac{k_B T}{\zeta_\perp^2} \sum_n q_n^4 \tau_n^2 \Bigg[a \cdot \frac{\dfrac{4\omega}{\tau_n}}{\omega^2 + \dfrac{4}{\tau_n^2}} + b \cdot \frac{4\omega\left(2/\tau_n + k_{\text{on}} + k_{\text{off}}\right)}{\omega^2 + \left(2/\tau_n + k_{\text{on}} + k_{\text{off}}\right)^2} \\
&\quad + c \cdot \frac{4\omega\left(\dfrac{1}{\tau_n} + k_{\text{on}} + k_{\text{off}}\right)}{\omega^2 + 4\left(1/\tau_n + k_{\text{on}} + k_{\text{off}}\right)^2} \Bigg],
\end{aligned} \qquad \text{(C.3)}
$$

with

$$
\begin{aligned}
a &= \left[1 + \frac{q_n^4 \bar{f}^2 k_{\text{off}}}{\zeta_\perp^2 N_0^2 k_{\text{on}}(k_{\text{on}} + k_{\text{off}})} \left(\tau_n - \frac{1}{2/\tau_n + k_{\text{on}} + k_{\text{off}}}\right)\right]^2 \\
b &= -\frac{q_n^4 \bar{f}^2 k_{\text{off}}}{\zeta_\perp^2 N_0 k_{\text{on}}(k_{\text{on}} + k_{\text{off}})} \cdot \frac{1}{2/\tau_n + k_{\text{on}} + k_{\text{off}}} \left[\frac{q_n^4 \bar{f}^2 k_{\text{off}}}{\zeta_\perp^2 N_0 k_{\text{on}}(k_{\text{on}} + k_{\text{off}})} \right. \\
&\quad \left. \cdot \left(\frac{1}{2/\tau_n + k_{\text{on}} + k_{\text{off}}} - \tau_n\right) + 1\right] \\
c &= \frac{q_n^8 \bar{f}^4 k_{\text{off}}^2}{\zeta_\perp^4 N_0^2 k_{\text{on}}^2 (k_{\text{on}} + k_{\text{off}})^2} \cdot \frac{1}{\left(2/\tau_n + k_{\text{on}} + k_{\text{off}}\right)^2}.
\end{aligned} \qquad \text{(C.4)}
$$

Inserting this into (C.1) we get

$$J'(\omega) = \frac{k_B T}{\pi \zeta_\perp^2} \sum_n q_n^4 \tau_r^2 \mathcal{P} \int_{-\infty}^{\infty} \frac{1}{\omega'-\omega} \left[\frac{\frac{4\omega' a}{\tau_n}}{\omega'^2 + \frac{4}{\tau_n^2}} + \frac{4\omega' b \left(\frac{2}{\tau_n} + k_{\text{on}} + k_{\text{off}}\right)}{\omega'^2 + \left(\frac{2}{\tau_n} + k_{\text{on}} + k_{\text{off}}\right)^2} \right. $$

$$\left. + \frac{4\omega' c \left(\frac{1}{\tau_n} + k_{\text{on}} + k_{\text{off}}\right)}{\omega'^2 + 4\left(\frac{1}{\tau_n} + k_{\text{on}} + k_{\text{off}}\right)^2} \right] d\omega' \qquad (C.5)$$

In order to solve this integral we need a partial fraction expansion. For the first term of the sum this means:

$$\frac{\frac{4\omega' a}{\tau_n}}{\left(\omega'^2 + \frac{4}{\tau_n^2}\right)(\omega'-\omega)} = \frac{A}{\omega'-\omega} + \frac{B\omega' + C}{\omega'^2 + \frac{4}{\tau_n^2}}, \qquad (C.6)$$

with

$$A = \frac{\frac{4\omega a}{\tau_n}}{\omega^2 + \frac{4}{\tau_n^2}}; \quad B = -\frac{\frac{4\omega a}{\tau_n}}{\omega^2 + \frac{4}{\tau_n^2}}; \quad C = \frac{\frac{16 a}{\tau_n^3}}{\omega^2 - \frac{4}{\tau_n^2}}$$

Integrating this expression gives:

$$\frac{\mathcal{P}}{\pi} \int_{-\infty}^{\infty} \frac{A}{\omega'-\omega} d\omega = \lim_{\Omega \to \infty} \lim_{\epsilon \to 0^+} \frac{1}{\pi} \left(\int_{-\Omega}^{\omega-\epsilon} \frac{A}{\omega'-\omega} d\omega + \int_{\omega+\epsilon}^{\Omega} \frac{A}{\omega'-\omega} d\omega \right)$$

$$= \lim_{\Omega \to \infty} \lim_{\epsilon \to 0^+} \frac{A}{\pi} \left(\left[\ln(\omega-\omega')\right]\Big|_{-\Omega}^{\omega-\epsilon} + \left[\ln(\omega'-\omega)\right]\Big|_{\omega+\epsilon}^{\Omega} \right)$$

$$= \lim_{\Omega \to \infty} \frac{A}{\pi} \left(\ln \frac{\Omega - \omega}{\Omega + \omega} \right)$$

$$= \lim_{\Omega \to \infty} \frac{A}{\pi} \left(\ln \frac{1 - \frac{\omega}{\Omega}}{1 + \frac{\omega}{\Omega}} \right)$$

$$= 0, \qquad (C.7)$$

125

C. Calculating the real part of the response function

$$\frac{1}{\pi}\int_{-\infty}^{\infty}\frac{B\omega'+C}{\omega'^2+\frac{4}{\tau_n^2}}d\omega = \underbrace{\frac{1}{\pi}\int_{-\infty}^{\infty}\frac{B\omega'}{\omega'^2+\frac{4}{\tau_n^2}}d\omega'}_{=0\ (\text{antisymmetric})} + \frac{1}{\pi}\int_{-\infty}^{\infty}\frac{C}{\omega'^2+\frac{4}{\tau_n^2}}d\omega'$$

$$= C\frac{\tau_n}{2\pi}[\arctan\omega']\Big|_{-\infty}^{\infty}$$

$$= C\frac{\tau_n}{2}. \tag{C.8}$$

Proceeding in an analogous way for the other terms of the sum in (C.5), we obtain the real part of the response function:

$$J'(\omega) = \frac{k_B T}{\zeta_\perp^2}\sum_n q_n^4 \tau_n^2 \Bigg\{ \frac{8}{\omega^2 \tau_n^2 + 4}\Bigg[1 + \frac{q_n^4 \bar{f}^2 k_{\text{off}}}{\zeta_\perp^2 N_0 k_{\text{on}}(k_{\text{on}}+k_{\text{off}})}\Bigg]$$

$$\cdot \Bigg(\tau_n - \frac{1}{2/\tau_n + k_{\text{on}} + k_{\text{off}}}\Bigg)\Bigg]^2 - \frac{4\big(2/\tau_n + k_{\text{on}} + k_{\text{off}}\big)}{\omega^2 + \big(2/\tau_n + k_{\text{on}} + k_{\text{off}}\big)^2}$$

$$\cdot \Bigg\{\frac{q_n^4 \bar{f} k_{\text{off}}}{\zeta_\perp^2 N_0 k_{\text{on}}(k_{\text{on}}+k_{\text{off}})}\Bigg[1 + \frac{q_n^4 \bar{f}^2 k_{\text{off}} \tau_n\big(1/\tau_n + k_{\text{on}} + k_{\text{off}}\big)}{\zeta_\perp^2 N_0 k_{\text{on}}(k_{\text{on}}+k_{\text{off}})\big(2/\tau_n + k_{\text{on}} + k_{\text{off}}\big)}\Bigg]\Bigg\}$$

$$+ \frac{4\big(1/\tau_n + k_{\text{on}} + k_{\text{off}}\big)^2}{\omega^2 + 4\big(\frac{1}{\tau_n} + k_{\text{on}} + k_{\text{off}}\big)^2} \cdot \frac{q_n^8 \bar{f}^4 k_{\text{off}}^2}{\zeta_\perp^4 N_0^2 k_{\text{on}}^2 (k_{\text{on}}+k_{\text{off}})^2}$$

$$\cdot \frac{1}{\big(2/\tau_n + k_{\text{on}} + k_{\text{off}}\big)^2}\Bigg\}. \tag{C.9}$$

Bibliography

[1] G. J. Baxter, R. A. Blythe, W. Croft, and A. J. McKane. Utterance selection model of language change. *Phys. Rev. E*, 73(4):046118, 2006.

[2] W. Croft. *Explaining Language Change: An Evolutionary Approach*. Longman, 2000.

[3] G. Deutscher. *The Unfolding of Language*. Metropolitan Books, 2005.

[4] G. J. Baxter, R. A. Blythe, W. Croft, and A. J. McKane. Modeling language change: An evaluation of Trudgill's theory of the emergence of New Zealand English. *Language Variation and Change*, 21:257–296, 2009.

[5] M. A. Nowak. Evolutionary biology of language. *Phil. Trans. R. Soc. B*, 355(1403):1615–22, 2000.

[6] V. Loreto and L. Steels. Social dynamics - emergence of language. *Nature Physics*, 3(11):758–760, 2007.

[7] M. D. Hauser, N Chomsky, and W T Fitch. The faculty of language: what is it, who has it, and how did it evolve? *Science*, 298(5598):1569–79, 2002.

[8] D. Crystal. *Language Death*. Cambridge University Press, 2000.

[9] C. Castellano, S. Fortunato, and V. Loreto. Statistical physics of social dynamics. *Rev. Mod. Phys.*, 81(2):591–646, 2009.

Bibliography

[10] R. V. Solé, B. Corominas-Murtra, and J. Fortuny. Diversity, competition, extinction: the ecophysics of language change. *J. R. Soc. Interface*, 7:1647–1664, 2010.

[11] M. Pagel. Human language as a culturally transmitted replicator. *Nat. Rev. Genet.*, 10(6):405–415, 2009.

[12] C. Schulze, D. Stauffer, and S. Wichmann. Birth, survival and death of languages by monte carlo simulation. *Commun. Comput. Phys*, 3(2):271–294, 2008.

[13] W. S.-Y. Wang and J. W. Minett. The invasion of language: emergence, change and death. *Trends Ecol. Evol.*, 20(5):263–269, 2005.

[14] M. A. Nowak, N. L. Komarova, and P. Niyogi. Computational and evolutionary aspects of language. *Nature*, 417(6889):611–617, 2002.

[15] A. Cangelosi and D. Parisi, editors. *Simulating the evolution of language*. Springer Verlag, 2002.

[16] S. Pinker and P. Bloom. Natural language and natural selection. *Behav. Brain. Sci.*, 13(4):707–726, 1990.

[17] M. A. Nowak. *Evolutionary dynamics: exploring the equations of life*. Harvard University Press, 2006.

[18] C. D. Yang. Internal and external forces in language change. *Language Variation and Change*, 12:231–250, 2000.

[19] J. R. Hurford. Biological evolution of the Saussurean sign as a component of the language acquisition device. *Lingua*, 77:187–222, 1989.

[20] M. Oliphant and J. Batali. Learning and the emergence of coordinated communication. *Center for Research on Language newsletter*, 11, 1996.

Bibliography

[21] M. A. Nowak, J. B. Plotkin, and D. C. Krakauer. The evolutionary language game. *J. Theor. Biol.*, 200(2):147–62, 1999.

[22] P. E. Trapa and M. A. Nowak. Nash equilibria for an evolutionary language game. *J. Math. Biol.*, 41(2):172–88, 2000.

[23] W. G. Mitchener and M. A. Nowak. Chaos and language. *Proc. R. Soc. B*, 271(1540):701–4, 2004.

[24] M. A. Nowak, N. L. Komarova, and P. Niyogi. Evolution of universal grammar. *Science*, 291(5501):114–118, 2001.

[25] M. A. Nowak and D. C. Krakauer. The evolution of language. *Proc. Natl. Acad. Sci. USA*, 96:8028–8033, 1999.

[26] M. A. Nowak, D. C. Krakauer, and A. Dress. An error limit for the evolution of language. *Proc. R. Soc. B*, 266(1433):2131–2136, 1999.

[27] M. A. Nowak, J. B. Plotkin, and A. A. Jansen. The evolution of syntactic communication. *Nature*, 404:495–498, 2000.

[28] M. A. Nowak and N. L. Komarova. Towards an evolutionary theory of language. *Trends Cogn Sci (Regul Ed)*, 5(7):288–295, 2001.

[29] J. B. Plotkin and M. A. Nowak. Major transitions in language evolution. *Entropy*, 3:227–246, 2001.

[30] M. A. Nowak. The basic reproductive ratio of a word, the maximum size of a lexicon. *J. Theor. Biol.*, 204:179–189, 2000.

[31] N. L. Komarova, P. Niyogi, and M. A. Nowak. The evolutionary dynamics of grammar acquisition. *J. Theor. Biol.*, 209(1):43–59, 2001.

[32] M. Eigen and P. Schuster. Hypercycle – a principle of natural self-organization. *Naturwissenschaften*, 64(11):541–565, 1977.

Bibliography

[33] M. A. Nowak. From quasispecies to universal grammar. *Z. Phys. Chem.*, 216:5–20, 2002.

[34] J. Hofbauer and K. Sigmund. *Evolutionary games and population dynamics*. Cambridge University Press, 1998.

[35] N. L. Komarova and M. A. Nowak. Natural selection of the critical period for language acquisition. *Proc. Biol. Sci.*, 268(1472):1189–96, 2001.

[36] N. L. Komarova and M. A. Nowak. Language dynamics in finite populations. *J. Theor. Biol.*, 221(3):445–57, 2003.

[37] F. A. Matsen and M. A. Nowak. Win-stay, lose-shift in language learning from peers. *Proc. Natl. Acad. Sci. USA*, 101(52):18053–7, 2004.

[38] L. Steels. A self-organizing spatial vocabulary. *Artificial Life*, 2(3):319–32, 1995.

[39] J. Ke, J. Minett, C.-P. Au, and W. S.-Y. Wang. Self-organization and selection in the emergence of vocabulary. *Complexity*, 7:41, 2002.

[40] T. Lenaerts, B. Jansen, K. Tuyls, and B. De Vylder. The evolutionary language game: an orthogonal approach. *J. Theor. Biol.*, 235(4):566–82, 2005.

[41] A. Baronchelli, M. Felici, V. Loreto, E. Caglioti, and L. Steels. Sharp transition towards shared vocabularies in multi-agent systems. *J. Stat. Mech. – Theory E*, page P06014, 2006.

[42] L. Dall'Asta, A. Baronchelli, A. Barrat, and V. Loreto. Nonequilibrium dynamics of language games on complex networks. *Phys. Rev. E*, 74(3 Pt 2):036105, 2006.

Bibliography

[43] R. A. Blythe. Generic modes of consensus formation in stochastic language dynamics. *J. Stat. Mech.*, page P02059, 2009.

[44] P. L. Krapivsky, S. Redner, and E. Ben-Naim. *A Kinetic View of Statistical Physics*. Cambridge University Press, 2010.

[45] A. Baronchelli, L. Dall'Asta, A. Barrat, and V. Loreto. Nonequilibrium phase transition in negotiation dynamics. *Phys. Rev. E*, 76(5):051102, 2007.

[46] D. M. Abrams and S. H. Strogatz. Linguistics: modelling the dynamics of language death. *Nature*, 424(6951):900, 2003.

[47] L. L. Cavalli-Sforza, M. W. Feldman, K. H. Chen, and S. M. Dornbusch. Theory and observation in culturaltransmission. *Science*, 218:19–27, 1982.

[48] D. Nettle. Using Social Impact Theory to simulate language change. *Lingua*, 108:95–117, 1999.

[49] D. Nettle. Is the rate of linguistic change constant? *Lingua*, 108:119–136, 1999.

[50] J. Ke, T. Gong, and W. S.-Y. Wang. Language change and social networks. *Commun. Comput. Phys.*, 3(4):935–949, 2008

[51] S. Kirby. Spontaneous evolution of linguistic structure - an iterated learning model of the emergence of regularity and irregularity. *IEEE T. Evolut. Comput.*, 5(2):102–110, 2001.

[52] K. Smith, H. Brighton, and S. Kirby. Complex systems in language evolution: the cultural emergence of compositional structure. *Advances in Complex Systems*, 6:537, 2003.

Bibliography

[53] P. Niyogi and R. C. Berwick. The proper treatment of language acquisition and change in a population setting. *Proc. Natl. Acad. Sci. USA*, 106(25):10124–10129, 2009.

[54] T. L. Griffiths and M. L. Kalish. Language evolution by iterated learning with bayesian agents. *Cognitive Science*, 31(3):441–80, 2007.

[55] F. Reali and T. L. Griffiths. Words as alleles: connecting language evolution with bayesian learners to models of genetic drift. *Proc. R. Soc. B*, 277(1680):429–36, 2010.

[56] E. Lieberman, J.-B. Michel, J. Jackson, T. Tang, and M. A. Nowak. Quantifying the evolutionary dynamics of language. *Nature*, 449(7163):713–6, 2007.

[57] M. Pagel, Q. D. Atkinson, and A. Meade. Frequency of word-use predicts rates of lexical evolution throughout Indo-European history. *Nature*, 449(7163):717–20, 2007.

[58] E. G. Altmann, J. B. Pierrehumbert, and A.E. Motter. Niche as a determinant of word fate in online groups. *PLoS ONE*, 6(5):e19009, 2011.

[59] R. Ferrer i Cancho and R. V. Solé. Two regimes in the frequency of words and the origins of complex lexicons: Zipf's law revisited. *Journal of Quantitative Linguistics*, 8(3):165–173, 2001.

[60] R. Ferrer i Cancho and R. V. Solé. Zipf's law and random texts. *Advances in Complex Systems*, 5(1):1–6, 2002.

[61] R. Ferrer i Cancho and R. V. Solé. Least effort and the origins of scaling in human language. *Proc. Natl. Acad. Sci. USA*, 100(3):788–91, 2003.

Bibliography

[62] R. V. Solé. Scaling laws in language evolution. In Claudio Cioffi-Revilla, editor, *Power Laws in the Social Sciences*. Cambridge University Press, 2006.

[63] M. Krauss. The world's languages in crisis. *Language* 68:4–10, 1992.

[64] W. J. Sutherland. Parallel extinction risk and global distribution of languages and species. *Nature*, 423:276–279, 2003.

[65] D Graddol. The future of language. *Science*, 303(5662):1329–1331, 2004.

[66] S. S. Mufwene. Language birth and death. *Annu. Rev Anthropol.*, 33:201–222, 2004.

[67] M. Patriarca and T. Leppänen. Modeling language competition. *Physica A*, 338(1–2):296–299, 2004.

[68] M. Patriarca and E. Heinsalu. Influence of geography on language competition. *Physica A*, 388(2-3):174–186, 2009.

[69] J. P. Pinasco and L. Romanelli. Coexistence of languages is possible. *Physica A*, 361(1):355–360, 2006.

[70] J. Mira and A. Paredes. Interlinguistic similarity and language death dynamics. *Europhys. Lett.*, 69(6):1031–1034, 2005.

[71] J. W. Minett and W. S.-Y. Wang. Modelling endangered languages: The effects of bilingualism and social structure. *Lingua*, 118:19–45, 2008.

[72] L. Chapel, X. Castello, C. Bernard, G. Deffuant, V. M. Eguiluz, S. Martin, and M. San Miguel. Viability and resilience of languages in competition. *PloS ONE*, 5(1):e8681, 2010.

Bibliography

[73] D. Stauffer, X. Castello, V. M. Eguiluz, and M. San Miguel. Microscopic abrams-strogatz model of language competition. *Physica A*, 374(2):835–842, 2007.

[74] X. Castelló, V. M. Eguiluz, and M. San Miguel. Ordering dynamics with two non-excluding options: bilingualism in language competition. *New J. Phys.*, 8:308, 2006.

[75] F. Vazquez, X. Castello, and M. San Miguel. Agent based models of language competition: Macroscopic descriptions and order-disorder transitions. *J. Stat. Mech.*, page P04007, 2010.

[76] X. Castello, V. M. Eguiluz, M. San Miguel, R. Toivonen, J. Saramäki, and K. Kaski. Modelling language competition: bilingualism and complex social networks. In *In A. Smith et al. (Eds), The Evolution of Language (EVOLANG7)*, pages 59–66. World Scientific, 2008.

[77] C. Schulze and D. Stauffer. Monte Carlo simulation of the rise and the fall of languages. *Int. J. Mod. Phys. C*, 16(5):781–787, 2005.

[78] D. Stauffer, C. Schulze, F. W. S. Lima, S. Wichmann, and S. Solomon. Non-equilibrium and irreversible simulation of competition among languages. *Physica A*, 371(2):719–724, 2006.

[79] D. H. Zanette. Analytical approach to bit-string models of language evolution. *Int. J. Mod. Phys. C*, 4:569–581, 2008.

[80] D. H. Zanette. Demographic growth and the distribution of language sizes. *Int. J. Mod. Phys. C*, 19(2):237–247, 2008.

[81] V. M. de Oliveira, M. A. F Gomes, and I. R. Tsang. Theoretical model for the evolution of the linguistic diversity. *Physica A*, 361(1):361–370, 2006.

[82] M. C. Corballis. Mirror neurons and the evolution of language. *Brain & Language*, 112(1):25–35, 2010.

[83] P. Lieberman, E. S. Crelin, and D. H. Klatt. Phonetic ability and related anatomy of the newborn and adult human, Neanderthal man, and the chimpanzee. *American Anthropologist*, 74(3):287–307, 1972.

[84] S. Pinker, M. A. Nowak, and J. J. Lee. The logic of indirect speech. *Proc. Natl. Acad. Sci. USA*, 105(3):833–8, 2008.

[85] J.-B. Michel, Y. K. Shen, A. Presser Aiden, A. Veres, M. K. Gray, Google Books Team, J. P. Pickett, D. Hoiberg, D. Clancy, P. Norvig, J. Orwant, S. Pinker, M. A. Nowak, and E. Lieberman Aiden. Quantitative analysis of culture using millions of digitized books. *Science*, 331(6014):176–82, Jan 2011.

[86] Google ngram viewer. http://books.google.com/ngrams.

[87] C. W. Gardiner. *Handbook of Stochastic Methods*. Springer, 2002.

[88] G. J. Baxter, R. A. Blythe, and A. J. McKane. Exact solution of the multi-allelic diffusion model. *Math. Biosci.*, 209(1):124–70, 2007.

[89] M. Newman. Power laws, Pareto distributions and Zipf's law. *Contemporary Physics*, 46:323–351, 2005.

[90] G. J. Baxter. The effect of population density on the propagation of language changes. In *Eighteenth Annual Colloquium of the Spatial Information Research Centre*, pages 99–105, 2006.

[91] G. J. Baxter, R. A. Blythe, and A. J. McKane. Fixation and consensus times on a network: A unified approach. *Phys. Rev Lett.*, 101:258701, 2008.

Bibliography

[92] F. Peruani and L. Tabourier. Directedness of information flow in mobile phone communication networks. *PLoS ONE*, 6(12), 2011.

[93] G. Palla, I. Derenyi, I. Farkas, and T. Vicsek. Uncovering the overlapping community structure of complex networks in nature and society. *Nature*, 435(7043):814–818, 2005.

[94] G. Palla, A.-L. Barabási, and T. Vicsek. Quantifying social group evolution. *Nature*, 446(7136):664–667, 2007.

[95] A. Lancichinetti, S. Fortunato, and J. Kertesz. Detecting the overlapping and hierarchical community structure in complex networks. *New J. Phys.*, 11:033015, 2009.

[96] G. Iniguez, J. Kertesz, K. K. Kaski, and R. A. Barrio. Opinion and community formation in coevolving networks. *Phys. Rev. E*, 80(6):066119, 2009.

[97] R. A. Blythe. The propagation of a cultural or biological trait by neutral genetic drift in a subdivided population. *Theor. Pop. Biol.*, 71(4):454–472, 2006.

[98] C. M. Grinstead and J. L. Snell. *Grinstead and Snell's Introduction to Probability*. American Mathematical Society, 2003.

[99] A. Singer, Z. Schuss, and D. Holcman. Narrow escape and leakage of Brownian particles. *Phys. Rev. E*, 78(5 Pt 1):051111, 2008.

[100] A. Singer, Z. Schuss, and D. Holcman. Narrow escape, part III: Nonsmooth domains and Riemann surfaces. *J. Stat. Phys.*, 122(3):491–509, 2006.

[101] S. Redner. *A Guide to First-Passage Processes*. Cambridge University Press, 2001.

: Bibliography

[102] J. T. Cox. Coalescing random walks and voter model consensus times on the torus in $\mathbb{Z}^d$. *Ann. Prob.*, 17(4):1333–1366, 1989.

[103] M. Davies. The corpus of contemporary american english as the first reliable monitor corpus of english. *Literary and Linguistic Computing*, 2007.

[104] B. New, M. Brysbaert, J. Veronis, and C. Pallier. The use of film subtitles to estimate word frequencies. *Applied Psycholinguistics*, 28:661–677, 2007.

[105] B. Alberts, D. Bray, J. Lewis, M. Raff, K. Roberts, and J. D. Watson. *Molecular Biology of the Cell*. Garland, 4th edition, 2002.

[106] E. Frey. Physics in cell biology: Actin as a model system for polymer physics. *Adv. in Solid State Phys.*, 41:345–356, 2001.

[107] J. F. Marko and E. D. Siggia. Stretching DNA. *Macromolecules*, 28:8759–8770, 1995.

[108] J. Käs, H. Strey, J. X. Tang, D. Finger, R. Ezzell, E. Sackmann, and P. A. Janmey. F-actin, a model polymer for semiflexible chains in dilute, semidilute, and liquid crystalline solutions. *Biophysical Journal*, 70:609–625, 1996.

[109] L. Le Goff, O. Hallatschek, E. Frey, and F. Amblard. Tracer studies on f-actin fluctuations. *Phys. Rev. Lett.*, 89(25):258101, 2002.

[110] K. Kroy and E. Frey. Force-extension relation and plateau modulus for wormlike chains. *Phys. Rev. Lett.*, 77(2):306–309, 1996.

[111] J. Wilhelm and E. Frey. Radial distribution function of semiflexible polymers. *Phys. Rev. Lett.*, 77(12):2581–2584, 1996.

Bibliography

[112] T. B. Liverpool and A. C. Maggs. Dynamic scattering from semiflexible polymers. *Macromolecules*, 34(17):6064–6073, 2001.

[113] O. Hallatschek, E. Frey, and K. Kroy. Tension dynamics in semiflexible polymers. I. coarse-grained equations of motion. *Phys. Rev. E*, 75:031905, 2007.

[114] O. Hallatschek, E. Frey, and K. Kroy. Tension dynamics in semiflexible polymers. II. scaling solutions and applications. *Phys. Rev. E*, 75:031906, 2007.

[115] B. S. Khatri and T. C. B. McLeish. Rouse Model with internal friction: A coarse grained framework for single biopolymer dynamics. *Macromolecules*, 40:6770–6777, 2007.

[116] A. R. Bausch and K. Kroy. A bottom-up approach to cell mechanics. *Nature Physics*, 2:231–238, 2006.

[117] O. Lieleg, M. M. A. E. Claessens, Y. Luan, and A. R. Bausch. Transient binding and dissipation in cross-linked actin networks. *Phys. Rev. Lett.*, 101(10):108101, 2008.

[118] O. Lieleg, K. M. Schmoller, M. M. A. E. Claessens, and A.R. Bausch. Cytoskeletal polymer networks: Viscoelastic properties are determined by the microscopic interaction potential of cross-links. *Biophys. J.*, 96(11):4725–4732, 2009.

[119] S. M. Volkmer Ward, A. Weins, M. R. Pollak, and D. A. Weitz. Dynamic viscoelasticity of actin cross-linked with wild-type and disease-causing mutant α-actinin-4. *Biophys. J.*, 95(10):4915–04923, 2008.

[120] T. B. Liverpool, M. C. Marchetti, J.-F. Joanny, and J. Prost. Mechanical response of active gels. *Europhys. Lett.*, 85:18007, 2009.

Bibliography

[121] C. Heussinger. Stress relaxation through crosslink unbinding in cytoskeletal networks. *New J. Phys.*, 14:095029, 2012.

[122] L. Wolff, P. Fernández, and K Kroy. Resolving the stiffening-softening paradox in cell mechanics. *PLoS ONE*, 7(7):e40063, 2012.

[123] O. Kratky and G. Porod. Röntgenuntersuchung gelöster Fadenmoleküle. *Rec. Trav. Chim. Pays-Bas.*, 68:1106–1123, 1949.

[124] N. Saitô, K. Takahashi, and Y. Yunoki. Statistical mechanical theory of stiff chains. *J. Phys. Soc. Jpn.*, page 219, 1967.

[125] O. Hallatschek. *Semiflexible Polymer Dynamics*. Shaker, 2004.

[126] M. Doi and S. F. Edwards. *The Theory of Polymer Dynamics*. Oxford University Press, 1986.

[127] G. C. Wick. The evaluation of the collision matrix. *Phys. Rev.*, 80(2):268–272, 1950.

[128] F. Gittes and F. C. MacKintosh. Dynamic shear modulus of a semiflexible polymer network. *Phys. Rev. E*, 58(2):1241–1244, 1998.

[129] J. S. Toll. Causality and the dispersion relation: Logical foundations. *Phys. Rev.*, 104(6):1760–1770, 1996.

[130] D. C. Morse. Viscoelasticity of tightly entangled solutions of semiflexible polymers. *Phys. Rev. E*, 58(2):1237–1240, 1998.

[131] F. C. MacKintosh, J. Käs, and P. A. Janmey. Elasticity of semiflexible biopolymer networks. *Phys. Rev. Lett.*, 75(24):4425–4428, 1995.

[132] T. G. Mason, T. Gisler, K. Kroy, E. Frey, and D. A. Weitz. Rheology of F-actin solutions determined from thermally driven tracer motion. *J. Rheol.*, 44(4):917–928, 2000.

Bibliography

[133] M. Rubinstein and R. H. Colby. *Polymer Physics*. Oxford University Press, 2003.

[134] T. Erdmann and U. S. Schwarz. Stochastic dynamics of adhesion clusters under shared constant force and with rebinding. *J. Chem. Phys.*, 121(18):8997–9017, 2004.

[135] T Erdmann and US Schwarz. Adhesion clusters under shared linear loading: A stochastic analysis. *Europhys. Lett.*, 66(4):603–609, 2004.

[136] U. Seifert. Rupture of multiple parallel molecular bonds under dynamic loading. *Phys. Rev. Lett.*, 84(12):2750–2753, 2000.

[137] P. Grassia and E. J. Hinch. Computer simulations of polymer chain relaxation via brownian motion. *J. Fluid Mech.*, 308:255–288, 1996.

[138] P. S. Lang. *Brownian Dynamics of Semiflexible Polymers*. Diploma thesis, Ludwig-Maximilians-Universität München, 2009.

[139] K. Kroy and J. Glaser. The glassy wormlike chain. *New J. Phys.*, 9:416, 2007.

[140] R. Albert and A. L. Barabási. Statistical mechanics of complex networks. *Rev. Mod. Phys.*, 74(1):47–97, 2002.

[141] M. Catanzaro, M. Boguna, and R. Pastor-Satorras. Generation of uncorrelated random scale-free networks. *Phys. Rev. E*, 71(2):027103, 2005.

[142] R. Tarjan. Depth-first search and linear graph algorithms. *SIAM Journal on Computing*, 1(2):146–160, 1972.

[143] D. T. Gillespie. General method for numerically simulating stochastic time evolution of coupled chemical reactions. *Journal of Computational Physics*, 22(4):403–434, 1976.

Bibliography

[144] D. T. Gillespie. Exact stochastic simulation of coupled chemical reactions. *J. Phys. Chem.*, 81(25):2340–2361, 1977.

[145] A. B. Bortz, M. H. Kalos, and J. L. Leibowitz. A new algorithm for Monte Carlo simulation of Ising spin systems. *Journal of Computational Physics*, 17:10–18, 1975.

i want morebooks!

Buy your books fast and straightforward online - at one of world's fastest growing online book stores! Environmentally sound due to Print-on-Demand technologies.

Buy your books online at

www.get-morebooks.com

Kaufen Sie Ihre Bücher schnell und unkompliziert online – auf einer der am schnellsten wachsenden Buchhandelsplattformen weltweit! Dank Print-On-Demand umwelt- und ressourcenschonend produziert.

Bücher schneller online kaufen

www.morebooks.de

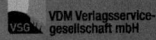
VDM Verlagsservicegesellschaft mbH
Heinrich-Böcking-Str. 6-8   Telefon: +49 681 3720 174   info@vdm-vsg.de
D - 66121 Saarbrücken   Telefax: +49 681 3720 1749   www.vdm-vsg.de

Printed by Books on Demand GmbH, Norderstedt / Germany